UNIVERSE

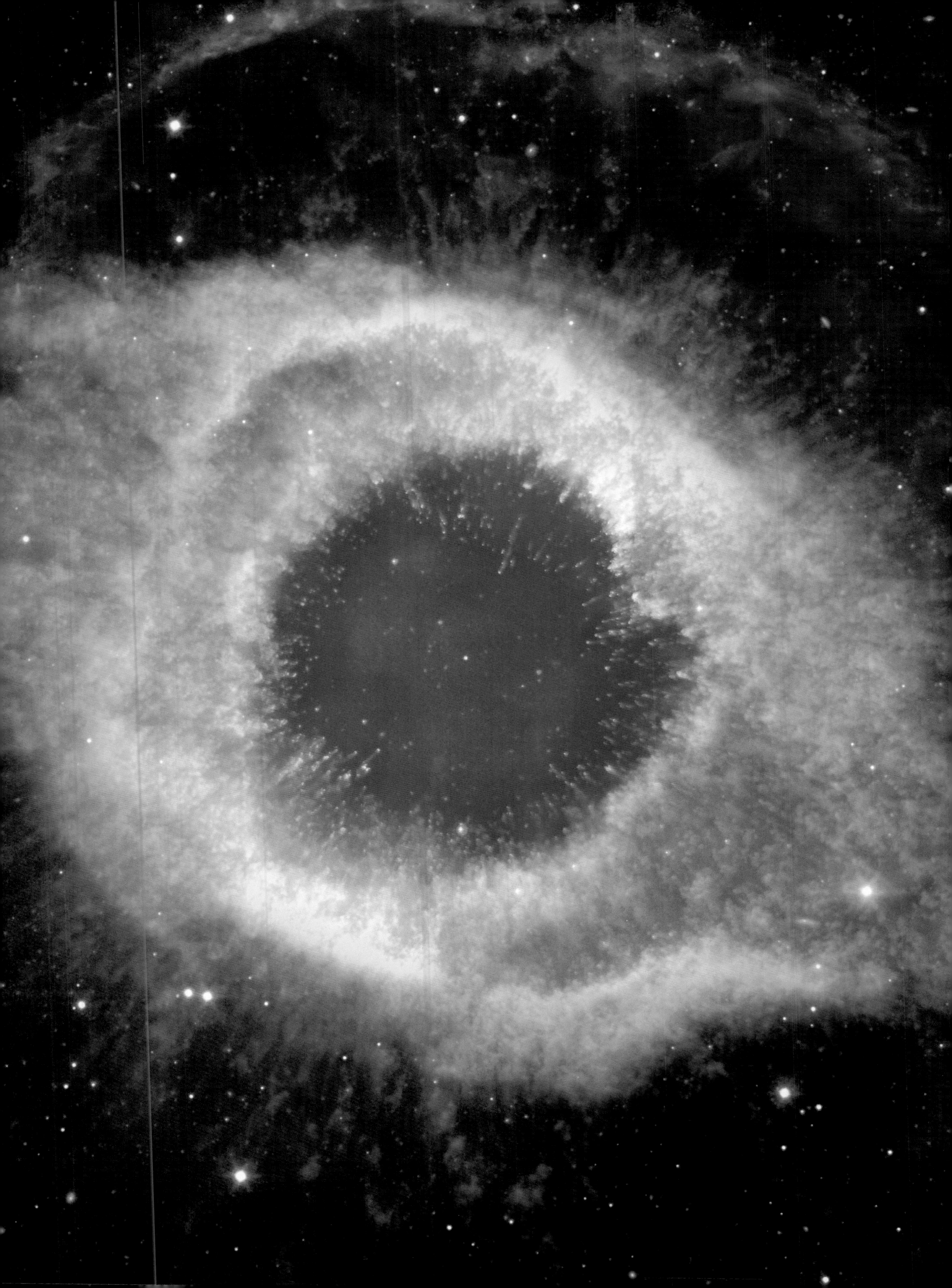

UNIVERSE

STUNNING SATELLITE IMAGERY FROM OUTER SPACE

HEATHER COUPER

FOREWORD BY

NIGEL HENBEST

THUNDER BAY
P·R·E·S·S
San Diego, California

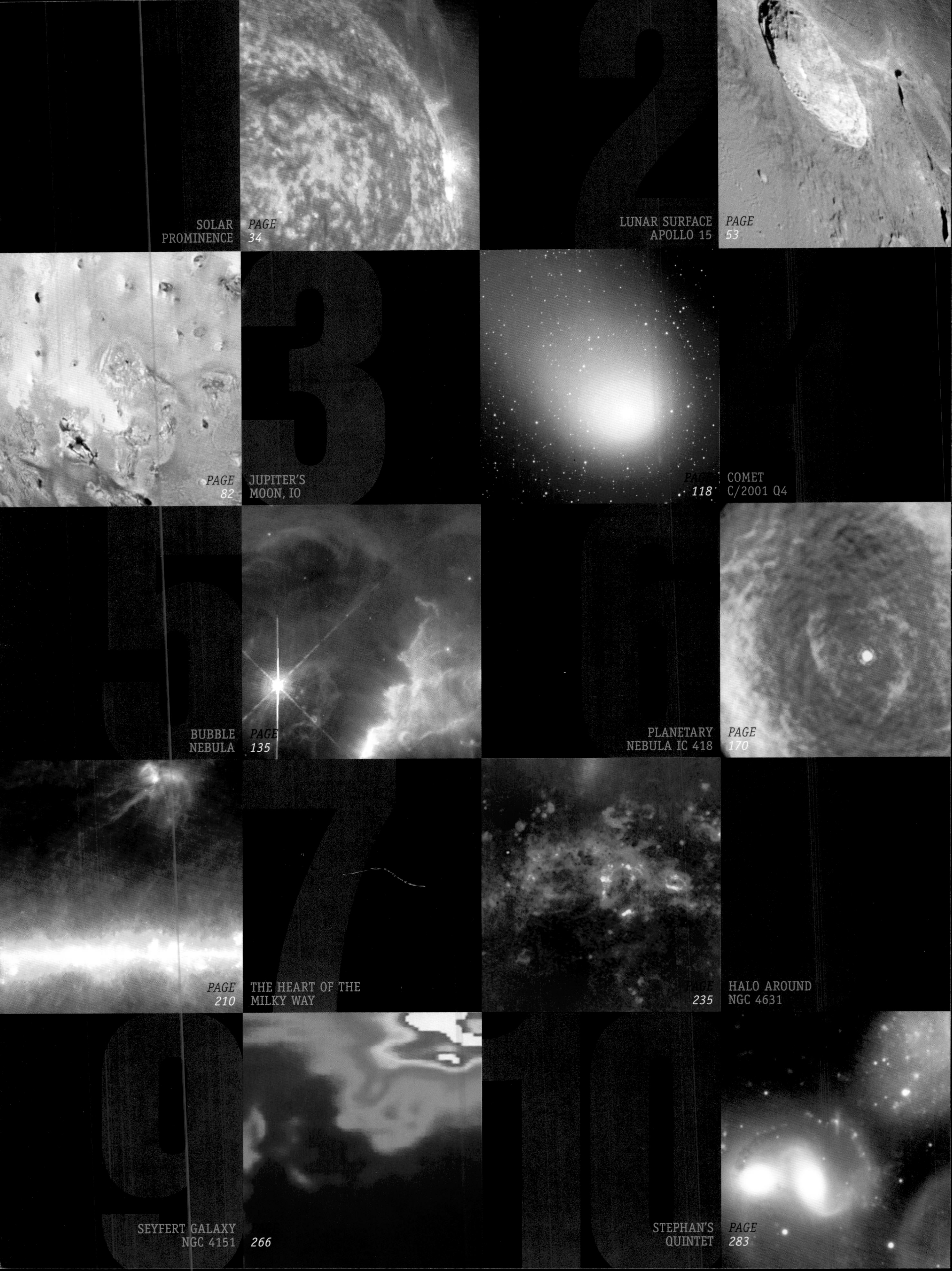
SOLAR PROMINENCE
PAGE 34
LUNAR SURFACE APOLLO 15
PAGE 53
PAGE 82
JUPITER'S MOON, IO
PAGE 118
COMET C/2001 Q4
BUBBLE NEBULA
PAGE 135
PLANETARY NEBULA IC 418
PAGE 170
PAGE 210
THE HEART OF THE MILKY WAY
PAGE 235
HALO AROUND NGC 4631
SEYFERT GALAXY NGC 4151
266
STEPHAN'S QUINTET
PAGE 283

CONTENTS

To John Pickup,
for wise counsel

FOREWORD

AS I LIFT MY GAZE FROM THE COMPUTER KEYBOARD, my eyes roam over some of the most wonderful landscapes in the world. England's Chiltern Hills are officially protected, as an Area of Outstanding Natural Beauty. In spring, the woods are thick with bluebells; in summer, magnificent red kites soar overhead borne on 5-foot/1.5m wingspans; in fall, the trees are as colorful as the woods of New England; winter brings glimmering slopes of snow.

Superb as it is, though, the landscape is only part of the view. As I look up, my eyes are filled with the other half: the skyscape.

The wonders of the skyscape start at home. The other day, a volcano erupted in Alaska, spewing dust high into the atmosphere and treating us to a magnificent salmon-pink sunset. At times an aurora spills natural pulsating disco lights across the heavens. A blood-red eclipse of the Moon is awesome, while the filigree of glowing tendrils surrounding the eclipsed Sun is literally breathtaking.

Take a small telescope out on a crystal-clear night, and you can peer deeper into the wonders of the skyscape. Start with the stark craters of the Moon and the glorious rings of Saturn, and travel out to the jewel-encrusted band of the Milky Way, bedecked with multicolored stars and glowing nebulae.

And over the past decades, we have been treated to the exploration of new "landscapes in the sky"—close-up views of other worlds as experienced by astronauts and by unmanned spacecraft.

Buzz Aldrin was one of the first two humans to see another world first hand. His awestruck description of the Moon perfectly sums up our companion in space: "Magnificent desolation."

Unmanned space probes have now visited all the major planets of our solar system—plus some oddments like comets and asteroids. Their postcards home have revealed fascinating, and beautiful, landscapes in space.

And new telescopes have treated us to a veritable orgy of new skyscapes far out in the depths of the cosmos. Chief among them is the Hubble Space Telescope, which from its perch above Earth's shimmering atmosphere has the clearest view of the universe. But it's far from being our only new eye on the sky. Keeping it company in orbit, Spitzer "sees" heat radiation from space while Chandra literally has X-ray vision.

These telescopes have provided a new vision of the universe. It's not just a quiet domain populated by well-behaved stars. Wreaths of gas bedeck the turbulent nebulae where stars are born. Clouds of roiling gas erupt from dying stars. Farther out in the cosmos, whole galaxies smash into one another in the mightiest traffic accidents of all. Black holes swallow stars whole and wreak havoc over millions of light years. And astronomers now have convincing proof that the whole universe was born in the ultimate act of violence: the big bang.

Along with its fellow satellites and colossal telescopes on the ground, Hubble was built for science. But its views are art. Hubble's images have literally opened the eyes of everyone on Earth to the wonders of the cosmos.

This magnificent book provides a gallery of the most stunning skyscapes we are privileged to see today—complete with an insightful commentary on each of these spectacular views. Enjoy!

AS A LITTLE KID OF SEVEN OR EIGHT, I was obsessed with the heavens. After all, my dad was an airline pilot, and he flew out of nearby Heathrow Airport in London. I was forever peeking through my curtains after bedtime, watching the international airplanes stacking over the airport, waiting to land.

One night of secret skygazing utterly changed my life. Across my view blazed a shooting star—my first—and it was green. I guiltily ran in to the living room to tell my parents, who were amused that I was up after hours. "Well, that's nice, dear," they observed. "But there's no such thing as a *green* shooting star."

On the front page of a national paper the next day was a small-but-perfectly-formed headline: "Green shooting star seen over West London."

The experience changed my whole existence. Gone was the ambition to follow in my father's footsteps as a pilot (just as well, in the interests of air safety): I was going to become an astronomer instead.

With my father's help, I learned the constellations of the night sky. With my mother's help, I learned to write down my experiences poetically. But one thing baffled me. I could see colors in the sky. My meteor had been green; the brilliant star Betelgeuse in Orion (and the planet Mars) were clearly red; while Rigel, the other dazzling star in Orion, was obviously blue-white. So why were the images in all the astronomy books shown in boring black-and-white?

Answer: the technology was not yet up to it. But, come the 1960s, everything changed.

I saw my first color pictures of the cosmos in an American magazine—and they blew me out of the water. In particular, the delicate purple-and-white traceries of the Veil Nebula—the remains of an exploded star—will forever haunt me.

Hence this book. It's a celebration of some of the most stunning, spectacular, and inspirational views of our universe. I defy anyone not to be moved by these images. But *who* can we thank for setting the ball in motion?

Astronomy has a long history, but we owe a phenomenal debt to Galileo, who in 1609 pointed the first astronomical telescope at the sky. The two-lensed "refractor" was borne of military technology: armies were using the combination of two lenses—one concave, one convex—to observe enemy camps at a distance.

Galileo sketched what he could see through his "optik tube." These included images of craters on the Moon; the phases of the planet Venus; and Jupiter's moon-system. His observations convinced him that the Earth was a mere planet, not the center of the universe. The religious authorities eventually confined him to house arrest for heresy.

Galileo's refractor was not perfect. By its very nature, it refracted (or bent) light—and light of different colors reacts differently, causing a kaleidoscopic haze around the object being observed. Isaac Newton—who, by coincidence, was born in the year that Galileo died—came up with a novel way to move forward. His design of a telescope—a reflector—collected starlight with a mirror instead of a lens. These "light buckets" are the key to how optical astronomy thrives today.

There's a limit to how big a lens can be before it sags (and distorts) under its own weight. But with mirrors, there isn't a problem. Soon—in the eighteenth century—William Herschel in Britain was building the biggest reflectors in the world. As well as discovering the planet Uranus, Herschel used his telescopes' light-grasp to explore our local star city, the Milky Way. He made a remarkably accurate sketch of its shape: but there was still no permanent, accurate way of recording astronomical observations.

Then, three incredible breakthroughs took place. By the nineteenth century, astronomers had telescopes that were sufficiently sensitive to measure distances to the stars. The invention of photography—in which William Herschel's son, John, was involved—allowed astronomers to record their observations in perpetuity. Finally, the pioneering British amateur astronomer William Huggins discovered spectroscopy—a way of splitting up the light from stars and planets to ascertain their composition.

The stage was now set for astronomy to move on. In the twentieth century, it changed from being a pursuit of gentlemen amateurs to become an in-your-face, cutting-edge science. The first hint of what was to come was the discovery of radio waves from space by Karl Jansky in the early 1930s. These observations of "cosmic interference"—exactly the same as the "snow" you see on your TV screen—led British scientist Stanley Hey in the 1940s to invent a new discipline: radio astronomy. Objects emitting radio waves are violent, destructive objects like black holes and the remains of exploded stars.

Hey's discovery was just the beginning of the "new astronomy." What you *see* of the universe is but a fragment: it's the middle chord on your piano. The radiation from the cosmos ranges from the bass notes of the radio waves to the ultimate descants of X-rays and gamma rays.

The whole plethora of information from the cosmos is reflected in this book. Observing the universe at different wavelengths tells us different tales about the physics of each source: how hot, how cold, how dense, how violent, and how different from our calm neighborhood in the solar system.

INTRODUCTION

And most of this multiwavelength cosmic symphony is only visible from space—above the Earth's blocking atmosphere. One astronomer commented that it's incredibly ironic that light, radio waves, X-rays, and ultraviolet radiation can travel to us across billions of light years of space—only to be cut off by a sixty-mile layer of gas.

The Hubble Space Telescope has exemplified the difference between being on-Earth and off-Earth. Seeing the heavens beneath our churning, restless atmosphere is akin to looking at the sky from beneath the depths of a swimming pool: above the turmoil, Hubble has a clear vision. And what a vision it has provided to humankind.

Many of the images you will see in this book have been captured in Earth orbit, or from spacecraft far beyond this that are examining the Sun, the planets, and the depths of the universe. These probes, which explore the whole architecture of the cosmos—at all wavelengths—are, quite literally, seeing our universe in a new light.

The images are actually subsidiary to the scientific data that modern-day astrophysicists collect. But romance is never far from the heart of astronomers. We are passionate about our subject, and we know that the *big* issues about the universe—its origin, the search for life, and the discovery of another Earth—are what drives us.

And the beautiful pictures we've collected encapsulate this philosophy. They are confirmation that we live in a truly fantastic cosmos.

Welcome to the Art Gallery of the Universe!

HEATHER COUPER

OUR SUN IS AN AVERAGE, MIDDLE-AGED, MIDDLE-CLASS STAR LIVING IN A SUBURB OF OUR MILKY WAY GALAXY. It might not sound like a very exciting proposition, but without the Sun, we would not exist. All life-forms on Earth, and possibly on other planets, crucially depend on it for energy, warmth, and light.

We have worshipped our nearest star for thousands of years. Religions have been based around it, temples have been constructed to celebrate it, and sacrifices—even human ones—have been made in its honor. Even today—in more cynical times—seeing a beautiful sunset is a hugely emotional experience.

The Sun was born in a cloud of gas 4.6 billion years ago. It had many siblings, which have since moved away on their own independent paths around our Milky Way. Many of these stars may well have given birth to planets, too. But a star is a very different entity from a planet. The Sun is made entirely of hydrogen gas. Because it is so massive (it could swallow over a million planets the size of the Earth), the pressure on the gas at its core forces it to undergo nuclear fusion reactions. At temperatures of 25.2 million degrees Fahrenheit/ 14 million degrees Celsius, the hydrogen nuclei are welded into helium—a process that generates prodigious quantities of energy—and keeps our star shining.

It takes nearly a million years for the energy from the core to worm its way to the Sun's surface. By then, temperatures have dropped to 9,932 degrees Fahrenheit/5,500 degrees Celsius—quite sufficient for the Earth to bask in its life-giving warmth and light.

The Sun's surface is a turbulent and constantly changing arena. It bubbles and foams as energy makes its way up from the core. In places, powerful magnetic fields damp down the activity, and our star breaks out in a rash of spots. Sunspots are cooler areas of the solar surface and appear dark against the brilliant background. Above them, crimson prominences—magnetic loops—arch into the Sun's upper atmosphere, the corona.

About every eleven years, the Sun bursts out in a frenzy of activity. Sunspots multiply, prominences explode into space, and huge solar flares eject electrically charged particles into space—which can cause chaos on our own planet by disrupting power lines and knocking out communications satellites.

But in the end, the Sun is an object of beauty. And this is particularly true of how it appears at the time of a total eclipse. By an amazing coincidence, the Moon and the Sun appear exactly the same size in our skies, and—when the time is right—the Moon can overlap the Sun exactly. That is when the faint, pearly corona flashes into view in the dark daytime sky and red prominences cling to the edges of the eclipsed disk. Experiencing an eclipse is awesome.

Our star is now halfway through its life cycle. In five billion years time, its nuclear reactor will shut down, having consumed all the hydrogen in its center. Briefly, the Sun will swell up to become a red giant star—possibly consuming the Earth as it expands—before ejecting its distended outer layers into space. All that will be left is a steadily cooling core, gradually merging into the blackness of space.

1 SUN

Dawn sky

OUR LOCAL STAR RISES IN THE EASTERN SKY, illuminating a vast complex of towering cloud formations. The beautiful, billowing cumulus clouds near the horizon are an indicator of a fine day to come. But the wispy, branching cirrus clouds—which are formed high in the upper atmosphere and are made of ice crystals—are a sign that unsettled weather is on the way. The Sun, as well as giving us light and warmth, drives our weather patterns. And the same is true for all the other planets in our solar system, too.

Previous pages | The Sun—our lifeline

OUR LIFE-GIVING SUN RISES OVER THE OCEAN, bathing the Earth in warmth and light. Like all the stars we see in the sky, it is a self-controlled nuclear reactor. It pours out energy by fusing hydrogen nuclei into helium—a process that makes our local star shine. Although the Sun burns up four million tons of hydrogen every second, it has enough resources to survive for at least five billion years into the future.

A rising sun

A BAOBAB TREE—*Adansonia digitata*—cradles the Sun as it rises over South Luangwa Park, Zambia. Baobabs, native only to Africa and India, have trunks that are only slightly smaller than the giant sequoia trees of North America. Our star is the driving force behind life on Earth, whether plant or animal. In the case of plants, such as a baobab tree, the Sun's light triggers a chemical process called photosynthesis. The energy in sunlight enables plants to absorb carbon dioxide from the atmosphere to help them flourish. They convert it into oxygen—which is essential to our survival.

Red sky at sunset

RED SKY AT NIGHT, SHEPHERD'S (OR SAILOR'S) DELIGHT? So goes the old adage. The reddening is caused by dust particles in the Earth's atmosphere. These remove all the short wavelengths of light that the Sun transmits and allow only the long-wavelength red to penetrate. Shepherds and sailors would be heartened by this image: a reddening, settled, dust layer on the horizon indicates that the atmosphere is calm, promising good weather for the next day. This evocative view was photographed at Big Sur, California.

Timelapse sunset

A VERY DIFFERENT VIEW OF SUNSET SEEN OVER THE OCEAN. Each image was captured at four-minute intervals. It shows that the Sun—at 8,700,000 miles/1.4 million km across—looks surprisingly small in our skies. But if you look at it near the horizon, as it is about to set, it looks bigger. This well-known optical illusion (usually called "the Moon Illusion") is caused by comparing the size of the Sun or Moon—when they are low in the sky—with nearby structures or clouds on Earth.

Sunset from space

ASTRONAUTS ON THE INTERNATIONAL SPACE STATION wax lyrical about the sunsets they see from orbit—and they get sixteen of them a day! "It casts an array of colors over what looks like a quarter of the planet," enthuses space doctor and astronaut Dave Rhys-Williams. Looking down on the Earth, they see the sunlight glowing through the different layers of our planet's atmosphere, beneath the darkness and vacuum of space. This image was captured by the crew of ISS Expedition One as they circled our world, 217 miles/350 km up, in 2001.

Sunrise from space ▷

A SUNRISE FROM SPACE, AS SEEN BY ASTRONAUTS on board a space shuttle mission. The arc is an effect caused by sunlight shining through the atmosphere. This image shows just how thin the Earth's atmosphere is—it extends a mere 62 miles/100 km above our planet's surface. Yet it protects us from the harsh conditions of space and from impacts by most meteoroids. Many space travelers comment on how their flights have woken them up to the vulnerability of the Earth.

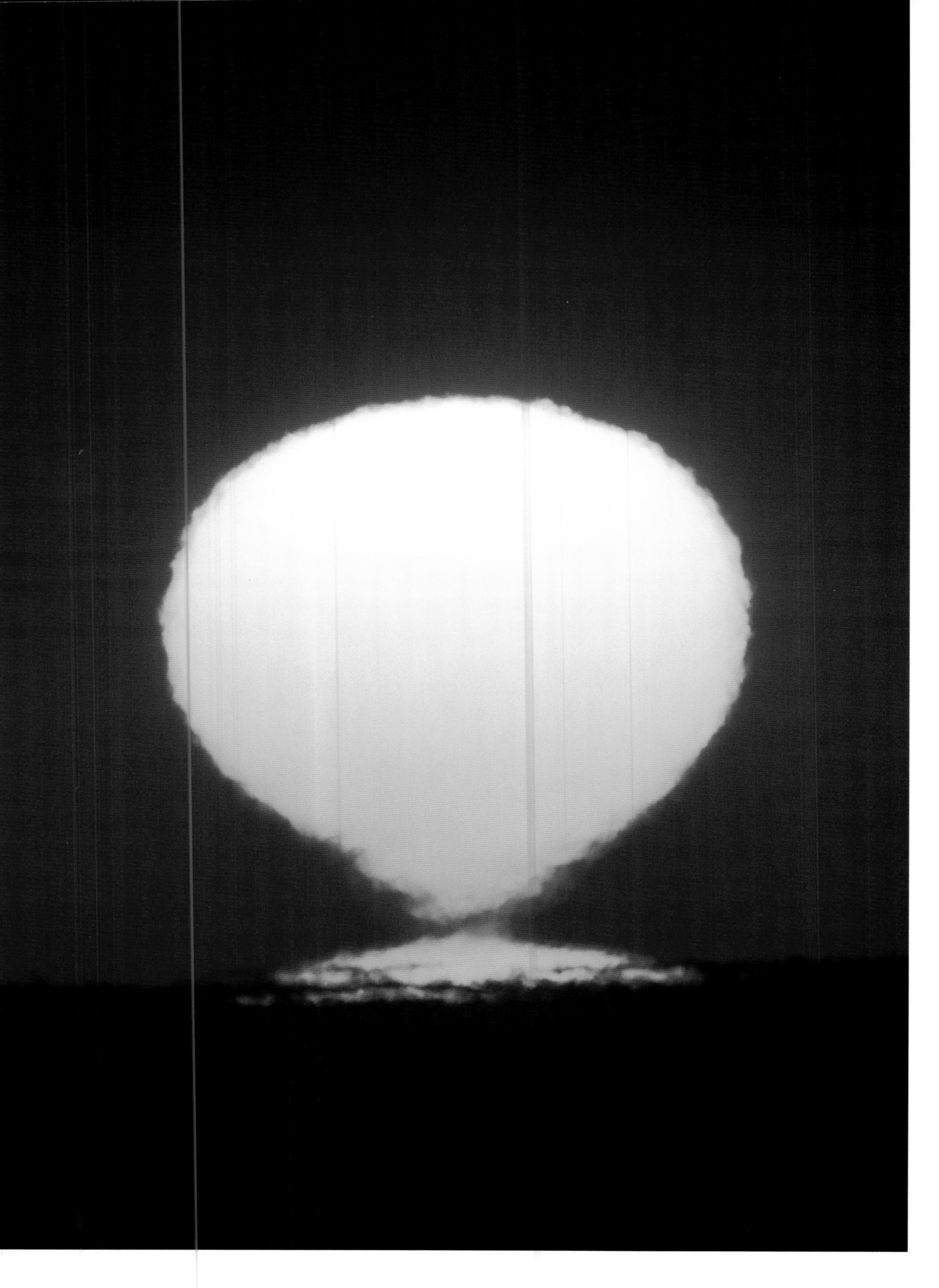

Distorted sun

AT SUNSET, THE SUN CAN TAKE ON SOME VERY STRANGE SHAPES. That is when you're looking at our local star close to the horizon and through many layers of the atmosphere. These refract sunlight and distort the Sun. On a very clear day, and with a flat horizon, the refraction can turn the last sliver of the setting Sun green: the best place to see this "Green Flash" is over a calm sea. In this image, a mirage—caused by sunlight refracted in a layer of hot air on the ground—sits below the distorted Sun.

Crepuscular rays

THE SUN GIVES RISE TO MANY GLORIOUS EFFECTS IN THE SKY, depending on weather conditions. For instance, solar halos and parhelia—"sun dogs"—are a sign that there are ice crystals in the atmosphere and that bad weather is on the way. In this image, the fingerlike crepuscular rays are caused by clouds below the horizon partly blocking off light from the setting Sun. The strong pink-purple sky is a result of dust in the atmosphere—often a result of a volcanic eruption.

Total solar eclipse

THE MOST AWE-INSPIRING SPECTACLE YOU CAN EVER SEE: a total eclipse of the Sun. They take place (about once every eighteen months) when the Moon exactly overlaps the Sun's disk. The Moon is four hundred times smaller than the Sun, but the Sun is four hundred times farther away—so the two objects appear the same size in the sky. You have to be in precisely the right part of the world to witness totality. It is only then that you can see the tendrils of the Sun's pearly corona—its beautiful, hot, outer atmosphere.

Diamond ring effect

AT THE END OF A TOTAL ECLIPSE, as the Moon moves away from the disk of the Sun, there is an unexpected bonus. A chink of sunlight shines through the Moon's irregular, mountainous edge, making the Sun look like a diamond ring in the sky. In this image, photographed from Aruba in the Caribbean in 1998, there is also a pink prominence at around the ten o'clock position. It is a gigantic loop of magnetized gas, connecting two sunpots.

Apollo 12 view of a solar eclipse

FLYING HOME FROM THE MOON ON APOLLO 12 IN 1969, astronaut Al Bean saw "a marvelous sight": "Our home planet eclipsed our local star," he recalls. This image shows the beginning of the eclipse, with sunlight refracted around the Earth by its atmosphere. Seen from the Moon, a total eclipse of the Sun by Earth would be an awesome sight. An astronaut there would see a dark Earth, with its cities twinkling like stars, move slowly and majestically in front of the Sun.

Timelapse eclipse ▷

ON NOVEMBER 3, 1994, there was a spectacular total eclipse visible from South America. Observers reported that, at the moment of totality, the insects stopped chirping. This timelapse image was captured high in the mountains of the Bolivian Andes. Photographed every fifteen minutes, the diminishing—and then afterward, the expanding—crescent of the Sun blazes its trail across the sky. The image at totality is overexposed: the darkness then makes the brighter planets visible in daytime. Brilliant Venus shines at top right; fainter Jupiter lies directly below.

◁ Timelapse annular eclipse

A SAGUARO CACTUS IN ARIZONA stands sentinel to an annular eclipse of the Sun. Taken in timelapse—every fifteen minutes—this image shows an almost-but-not-quite-total solar eclipse. Annular eclipses take place when the Moon, in its elliptical orbit, is just too far from the Earth to overlap the Sun completely. This eclipse took place on May 10, 1994.

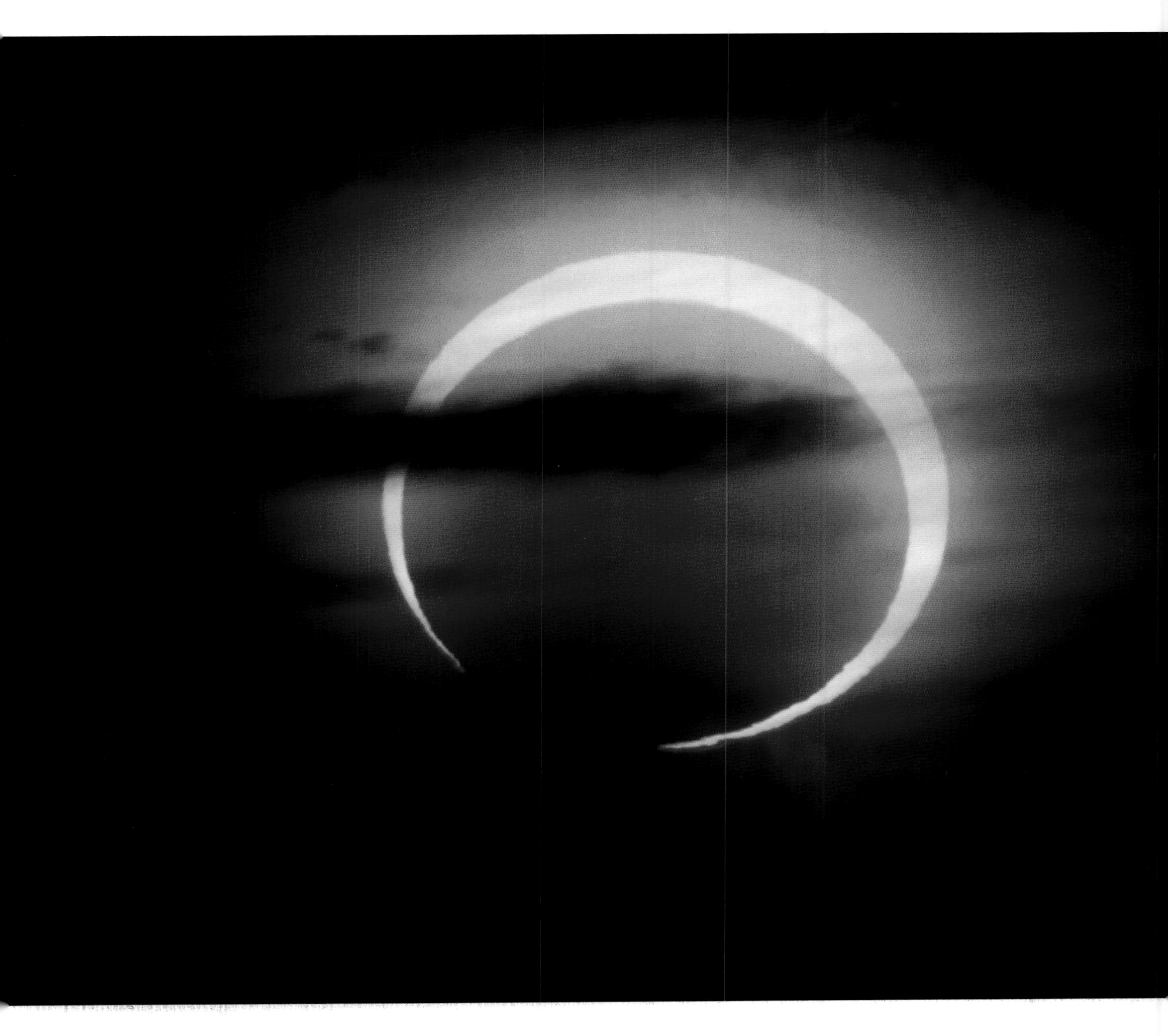

Annular solar eclipse

TAKEN NEAR SUNSET, WITH A RED SKY AND LOW CLOUDS, this photograph of an annular eclipse was captured in California on January 4, 1992. Because the Moon is too far from the Earth to cover the Sun, it reveals an "annulus" of light around the Sun—hence the name for this type of eclipse. The rim of the Sun is still so bright that it drowns out the much fainter prominences and the delicate corona. While annular eclipses are interesting, they are no substitute for the experience of totality.

◁ Hazy solar corona

IN MAY 1998, THE ORBITING SOLAR AND HELIOSPHERIC OBSERVATORY (SOHO) imaged the Sun's outer atmosphere—the corona—as our star approached maximum in its eleven-year cycle of magnetic activity. Taken in ultraviolet light, this color-coded view shows how hot the corona is, and the places where the magnetism is most intense (white regions). Blue areas are at a temperature of 1 million degrees Celsius/1.8 million degrees Fahrenheit; green regions are at 2.7 million degrees Fahrenheit/1.5 million degrees Celsius; red indicates temperatures of 3.6 million degrees Fahrenheit/2 million degrees Celsius.

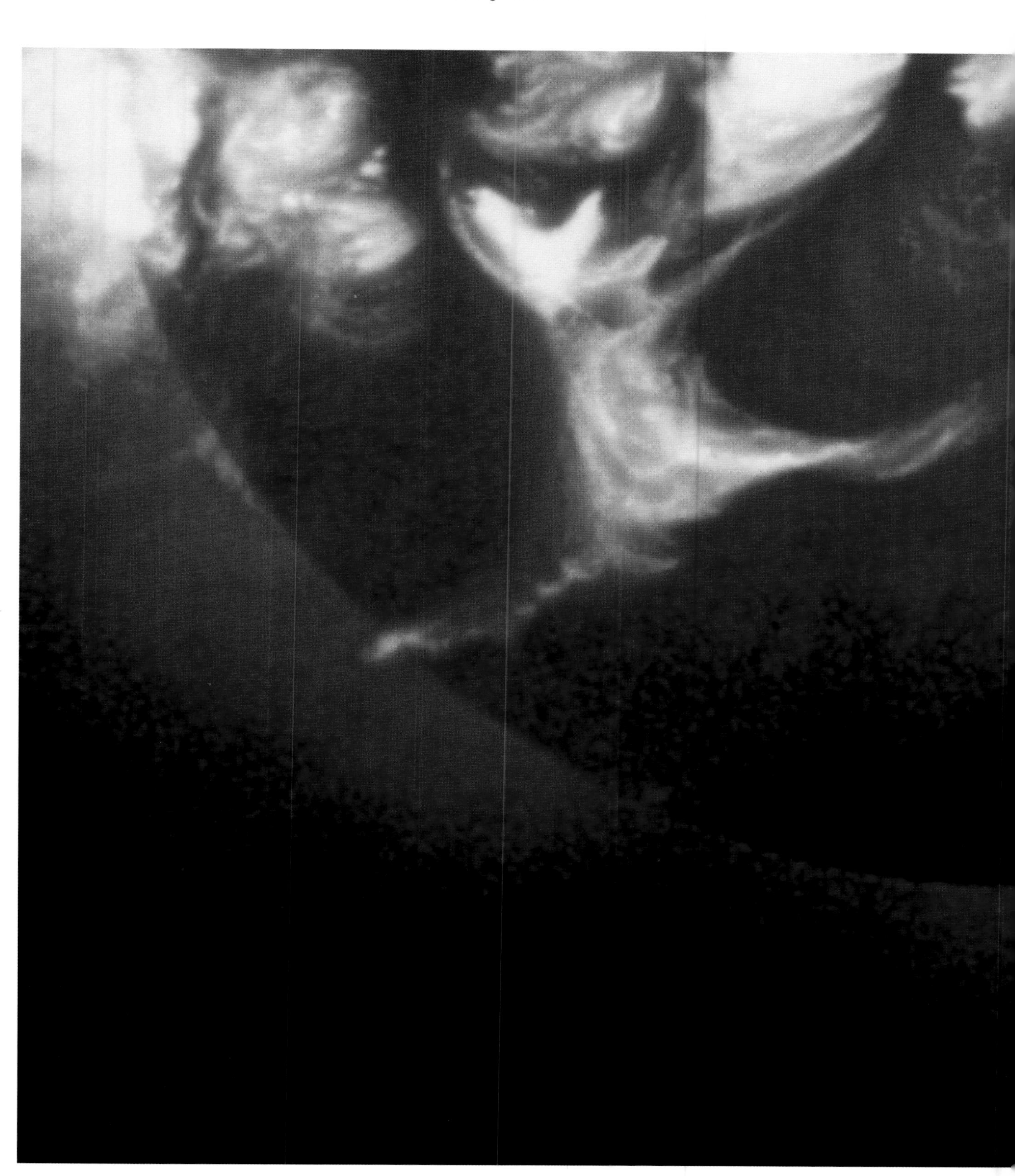

X-ray image of a filament in the solar corona

BEFORE THE ERA OF SATELLITES, astronomers had little knowledge of the Sun's corona. But this X-ray view—from Japan's YOHKOH satellite—reveals that it is a truly dynamic place. Here, in the center of the image, a prominence erupts, spewing electrically charged particles into space—particles that can affect telecommunications as far away as Earth. Prominences are vast magnetic loops that hover above groups of sunspots. This particular prominence delivered an enormous burst of energy—yet it lived for only a few hours.

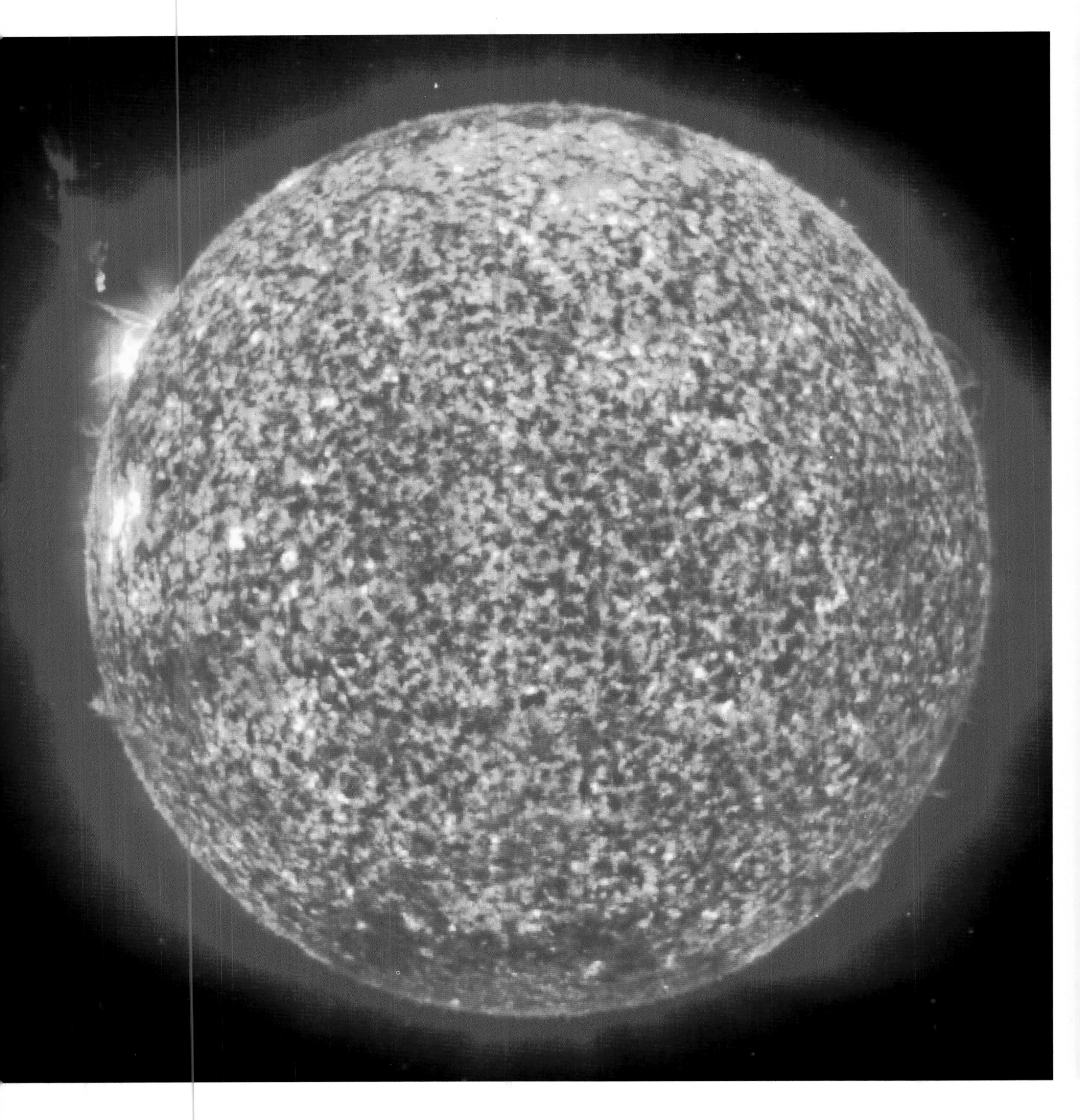

The Sun's atmosphere

THIS SOHO IMAGE SHOWS HOW THE SUN'S LOWER ATMOSPHERE would look through ultraviolet eyes. At a temperature of 108,000 degrees Fahrenheit/60,000 degrees Celsius, this region—which hovers above the Sun's surface—is called the chromosphere (literally, the "color sphere"), because of its redness. Turbulent gases make it appear mottled. And at the edge of the Sun, prominences stand out against our local star's limb. The prominence at top left is ten times the size of the Earth.

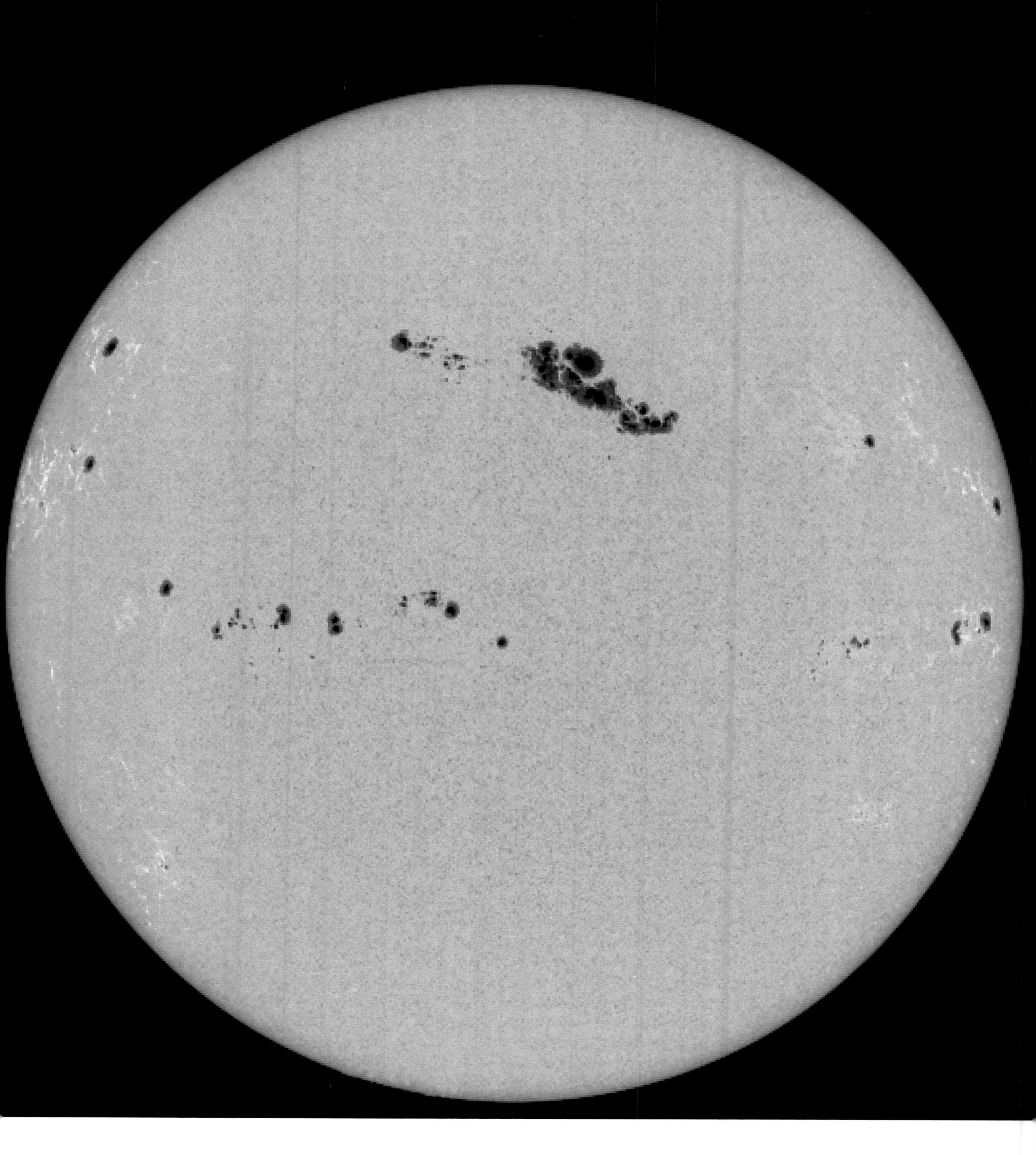

Sunspots

LIKE ALL STARS, OUR SUN IS SPOTTY. Sunspot activity waxes and wanes on a cycle of about eleven years—although the jury is still out as to exactly why it happens. What is certain is that magnetic fields from its interior block the flow of its hot, circulating gases and cool them down, so that sunspots appear as darker regions on its surface. But everything is relative. These sunspots are at a temperature of a "mere" 8,132 degrees Fahrenheit/4,500 degrees Celsius (as opposed to the Sun's surface temperature of 9,932 degrees Fahrenheit/5,500 degrees Celsius). The large sunspot group at the top is over ten times the width of our planet.

Sun surface

ALTHOUGH THE SUN DOESN'T HAVE A SOLID SURFACE—it is entirely made of superhot gas—its so-called photosphere is the outermost region of our local star's opaque gases. Here, in an image from the National Solar Observatory in New Mexico, we can see the structure of our star's "surface." Rising gases from within produce a bubbling structure, called "granulation," where currents of heat ascend and descend. Each cell on this photograph is as large as Texas.

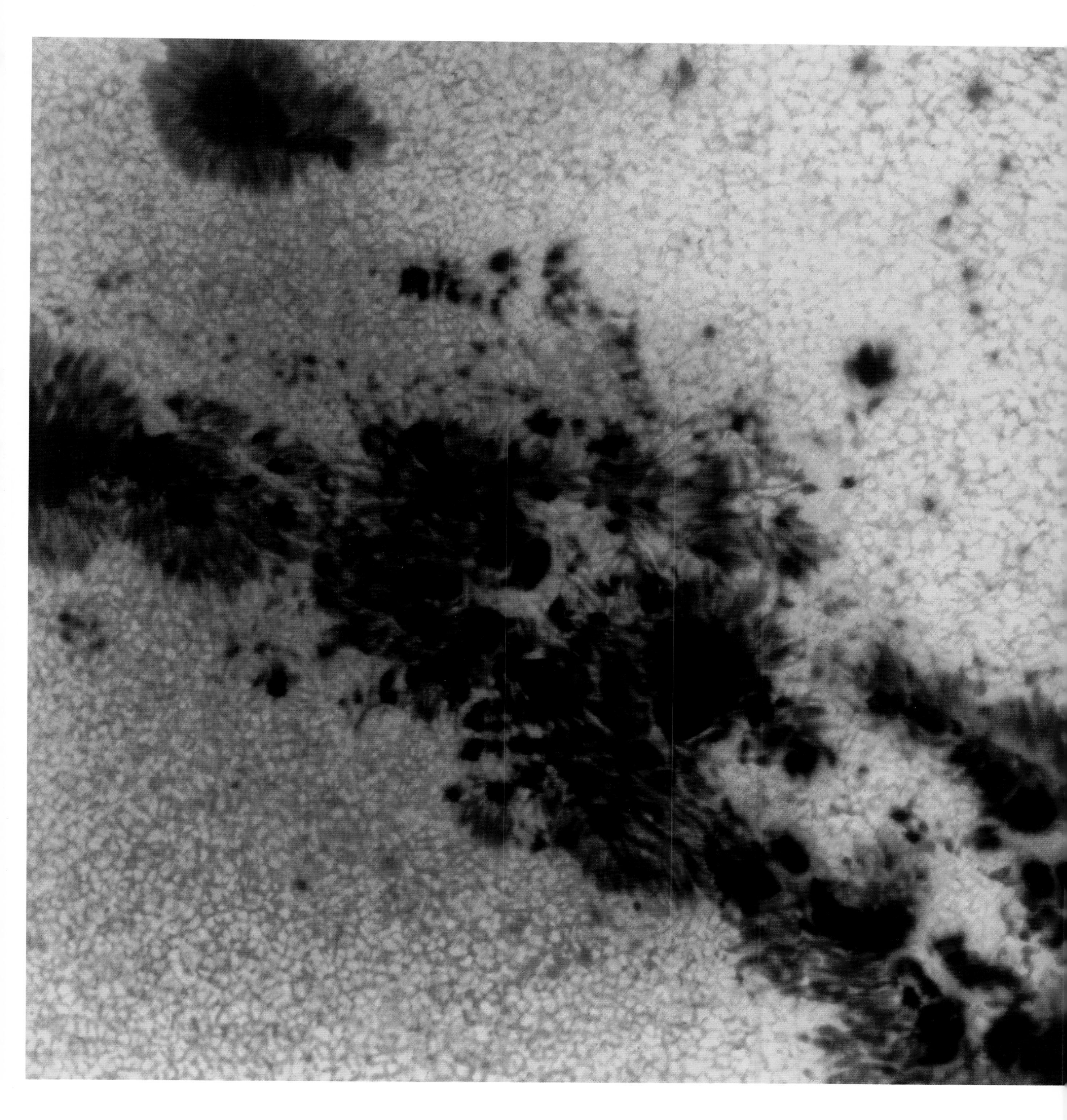

Sunspot group and granulation

CHINESE ASTRONOMERS IN ANCIENT TIMES were the first to observe sunspots with the unaided eye. If you have eclipse goggles, you can often pick out large spots on the Sun—but never look at the Sun directly, because the infrared light (heat radiation) could blind you. This image shows the structure of a large sunspot group. The cool "umbra," where the magnetic blocking is at its strongest, looks dark. The surrounding, lighter, "penumbra" is slightly warmer.

Plasma loops

OUR NEAREST STAR IS FOREVER ACTIVE. This image, captured in ultraviolet by satellite TRACE, reveals a plasma loop that highlights the magnetic field between two sunspots. The surface of the Sun appears dark in contrast, but the loop blazes brilliantly at a temperature of 1.8 million degrees Fahrenheit/1 million degrees Celsius. The cause of the Sun's magnetism is still under scrutiny—although experts believe that it could be because our gaseous star "winds itself up" by rotating at different speeds at different latitudes.

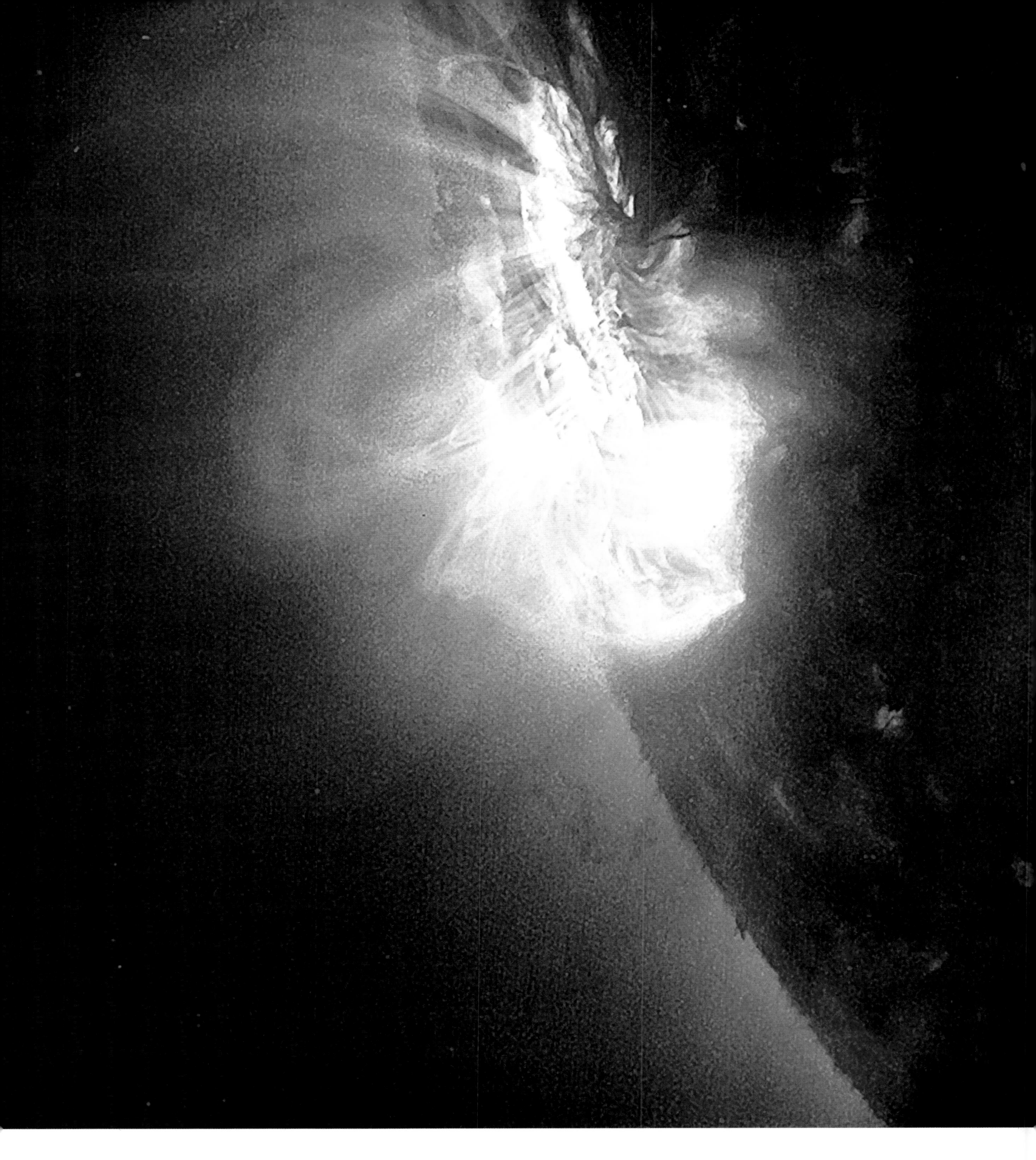

Colored TRACE image of plasma loops

THIS ENORMOUS PLASMA LOOP could swallow fifteen planets the size of Earth. Imaged by the TRACE (Transition Region and Coronal Explorer) satellite in 1998—just as the Sun was approaching solar maximum in its magnetic activity—it shows a spectacular outburst from a group of sunspots. The colors reveal the intense temperatures that are characteristic of a typical star. Blue represents 360,000 degrees Fahrenheit/ 200,000 degrees Celsius; green 900,000 degrees Fahrenheit/500,000 degrees Celsius; and red, 2.7 million degrees Fahrenheit/1.5 million degrees Celsius.

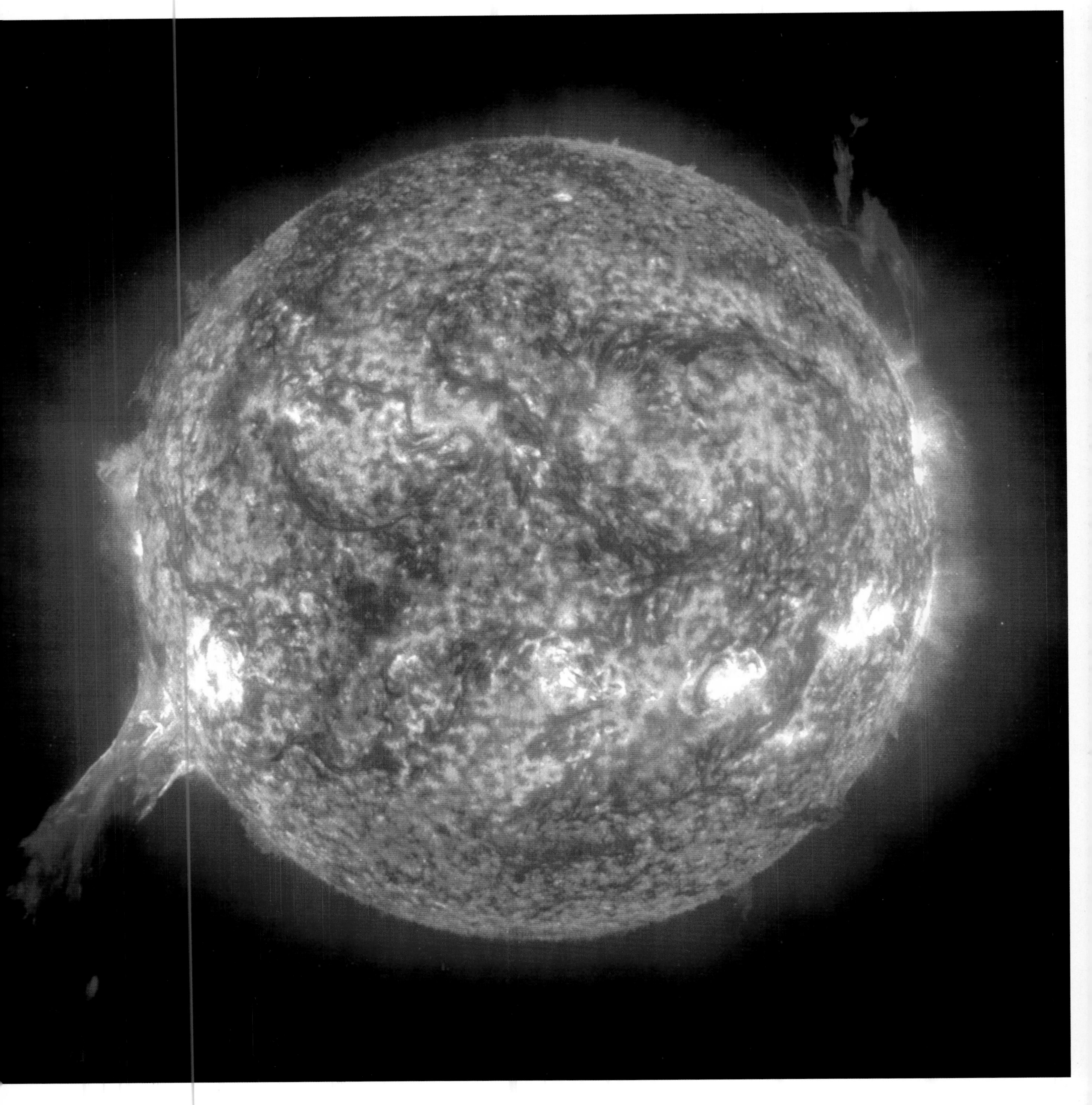

Solar prominence

THE SOHO SATELLITE CAPTURED THIS IMAGE of an erupting solar prominence (lower left). These outbursts are vast versions of plasma loops, resulting from interactions between widely spaced sunspot groups. Here, the Sun is near solar maximum, with most of its activity being confined to its equator (bright regions). Solar minimum starts with spots being formed near the Sun's poles—after which they gradually migrate toward our star's central regions.

Solar flare ▷

OCCASIONALLY, THE SUN CAUSES TROUBLE. That's when the lines of its complex magnetic field connect and then short-circuit—resulting in a spectacular explosion called a solar flare. This sequence of flare images spans a period of eight hours. The electrical particles that result from this mayhem reach Earth within hours, causing havoc. Stock exchanges have gone down, and nationwide power outages have frequently taken place. But there is a bonus. We get to see the particles hitting our atmosphere as a beautiful display of aurora.

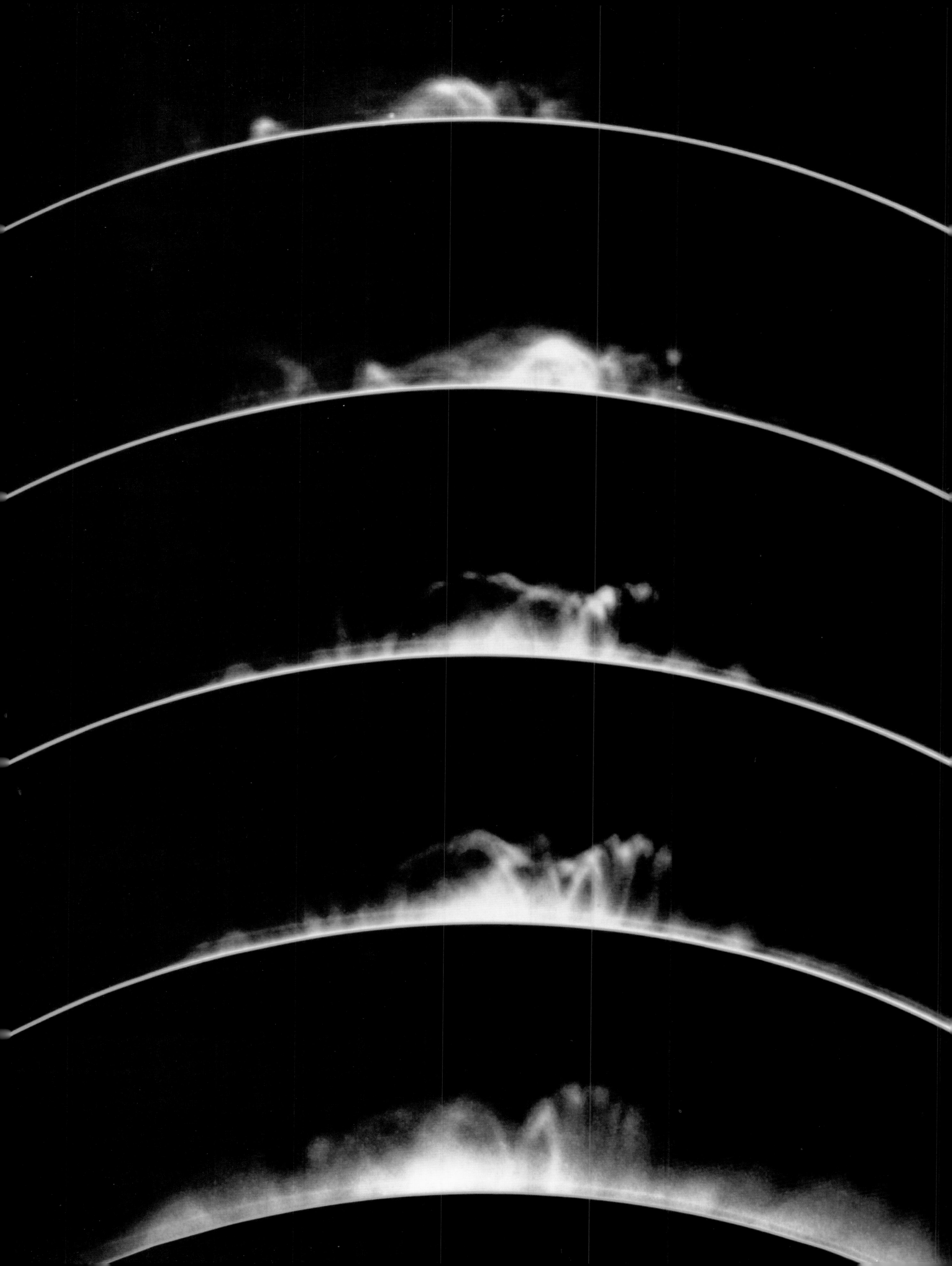

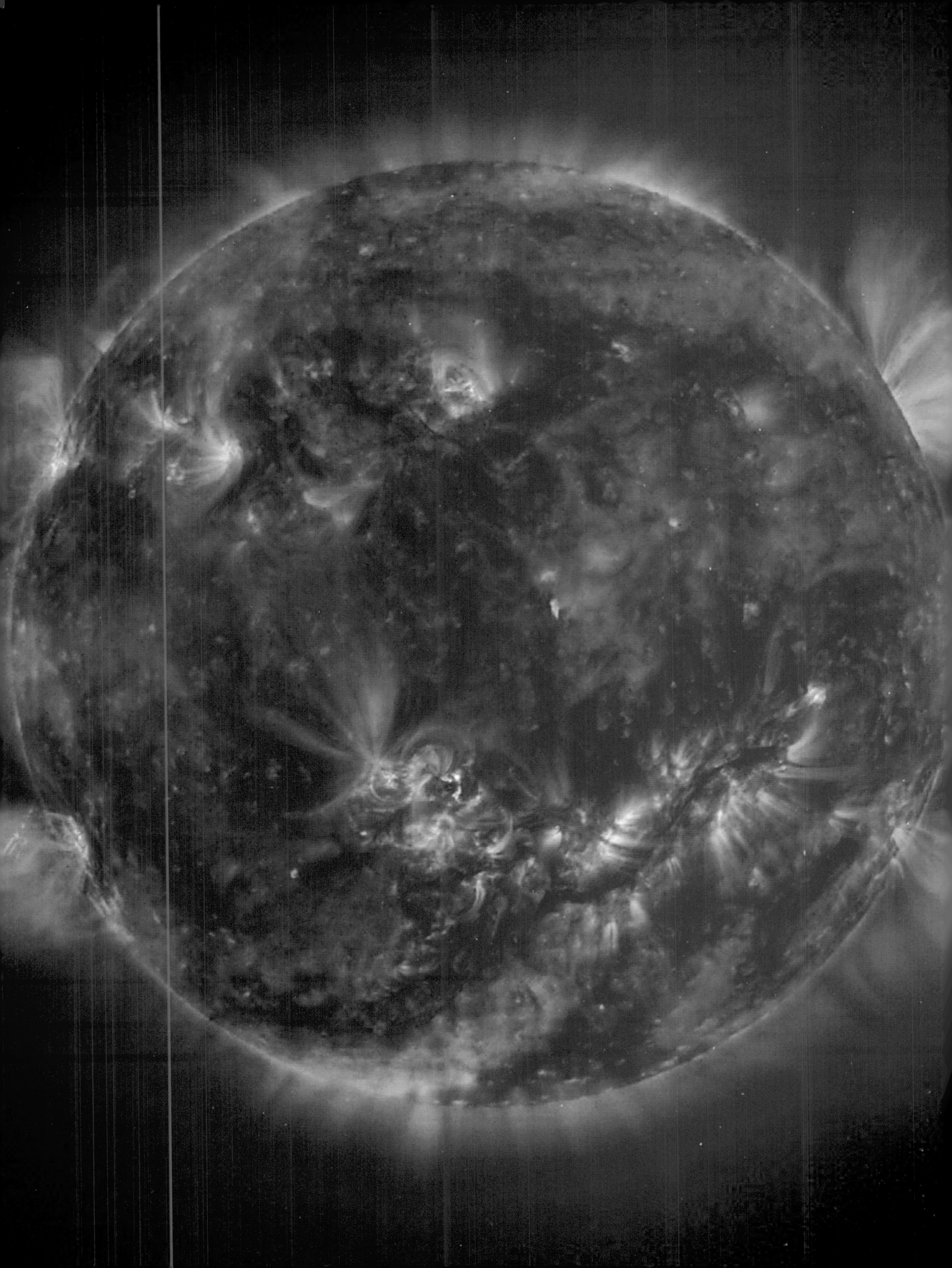

◁ Active Sun

THIS SOHO IMAGE SHOWS THE DELICATE TRACERY OF MAGNETIC LOOPS in the corona of our local star, where the temperature is around 1.8 million degrees Fahrenheit/1 million degrees Celsius. The corona is also the origin of the solar wind—a blast of electrically charged particles that are ejected from the Sun at speeds of up to 497 miles per second/800 km/s. The solar wind pervades the whole solar system, and it has effects on all the planets. Probes sent into space from Earth also need to contend with the solar wind, which whips itself up into a frenzy every eleven years.

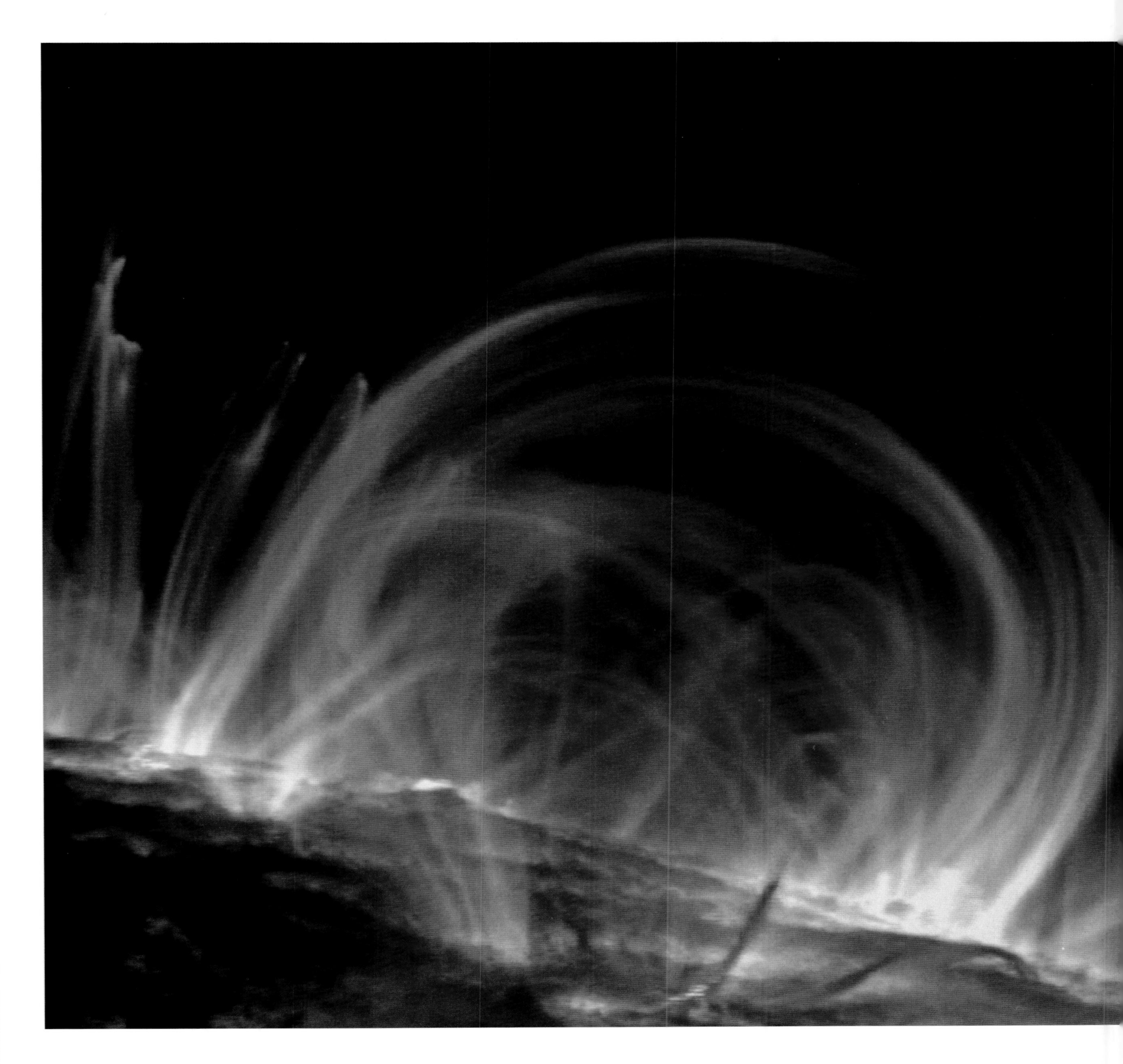

Coronal loops

HOW THE HOT CORONA GOT HOT: In this TRACE image, a coronal loop arches over the cooler surface of the Sun (shown dark). The hottest areas of the magnetic loops are at the bottom (shown white)—and they are capable of warming the Sun's atmosphere to thousands of times the temperature of its surface. The gas in the loop follows the path of the arched magnetic field and crashes back to the surface at a speed of over 62 miles per second/100 km/s.

O moon, when I gaze
on thy beautiful face,
Careering along through
the boundaries of space,
The thought has often
come into my mind
If ever I shall see thy
glorious behind.

THE NINETEENTH-CENTURY ENGLISH POET EDMUND GOSSE claimed that his housekeeper penned this immortal quatrain. Perhaps it isn't the best poetry, but it is an accurate observation. The Moon always presents the same face toward the Earth: the face of "the Man in the Moon"—venerated since antiquity by people around the world. The reason for this is gravity. The Earth and Moon are so close to each other—and the Moon so large—that our lunar companion exerts huge tides on the Earth's oceans. This grip on our world has come at the Moon's cost: the energy loss has caused it to stop spinning.

The "Man in the Moon" is made up of a series of features, gouged out by giant asteroids some 3.8 billion years ago, filled with dark lava that has welled up from the lunar interior. In fact, the whole of the Moon's surface is testimony to constant bombardment by meteoroids of all sizes. Craters are everywhere, some inside or overlapping others.

Although the Moon is large—2,160 miles/3,476 km across—it isn't massive enough to hold on to an atmosphere. Unprotected from the abuse hurled onto it from space, the Moon bears its scars blatantly. The lack of an atmosphere also means that the rate of erosion is slow: even the most ancient craters have a presence. In contrast, on Earth—which has a destroying atmosphere—craters are hard to find, although both bodies have been equally bombarded.

The brightest (and youngest) craters, like Aristarchus, are visible with the unaided eye. And the biggest craters on the Moon are truly awesome: Clavius (the largest) is 140 miles/225 km across.

The Moon circles our planet in 29.5 days, changing its shape from crescent to full—and then back again—as its dark surface is illuminated by the Sun. About twice a year, it passes into the Earth's shadow, and for about an hour, we witness a total lunar eclipse.

Lying 238,855 miles/384,400 km away from Earth, the Moon is a mere three-day journey through the blackness of space. Nevertheless, in 1961, it seemed impossible to think of humans going there. President John F. Kennedy's promise in a speech made that year—to put astronauts on the Moon within an astonishing eight years—was nothing short of audacious.

Although some dismiss the Apollo space missions as political posturing, we have undoubtedly learned a vast amount about the Moon as a result. We now know that our celestial companion originated as a result of "the Big Splash"—a collision between the Earth and a body the size of Mars. And we know that the Moon may contain limited amounts of water and exotic minerals, which could become a useful mining resource.

Now many more nations are planning a return to the Moon. It will become our new focus—and perhaps our new homeland—throughout the twenty-first century.

2 MOON

Previous pages | The Moon—our cosmic island

THE MOON—OUR COMPANION IN SPACE—is one of the reasons why we have become a space-faring community. Ancient Polynesian civilizations would travel by canoe to the next island—just because, in the words of mountaineer Sir Edmund Hillary, "it was there." And then they gained the courage to venture further afield. So it is in the case with the Moon: our first cosmic island. In this image, the full moon appears squashed and distorted because it is close to the horizon and seen though layers of air.

Crescent moon from Earth

PHOTOGRAPHED FROM NEW SOUTH WALES, AUSTRALIA, JUST AFTER SUNSET, the slender crescent moon hangs in the twilight sky. The Moon has only recently emerged from being almost in line with the Sun during its monthly orbit (the word "month" probably derives from "moon")—so only a sliver of it is illuminated. Northern-hemisphere dwellers will be surprised at the configuration: the waxing Moon south of the equator is back-to-front compared to the view they see. The planet on the left is brilliant Venus.

Full Moon from Earth ▷

A SEA ARCH AT CABO SAN LUCAS IN BAJA CALIFORNIA, MEXICO, provides a dramatic backdrop for this image of a rising full moon. The Moon is full when it is directly opposite the Sun in the sky, reflecting our star's light directly toward the Earth. Many cultures have variously interpreted the dark markings on the full moon as a face, a rabbit, or a beautiful woman. In fact, the dark markings are huge craters caused by a massive bombardment from asteroids 3.8 billion years ago.

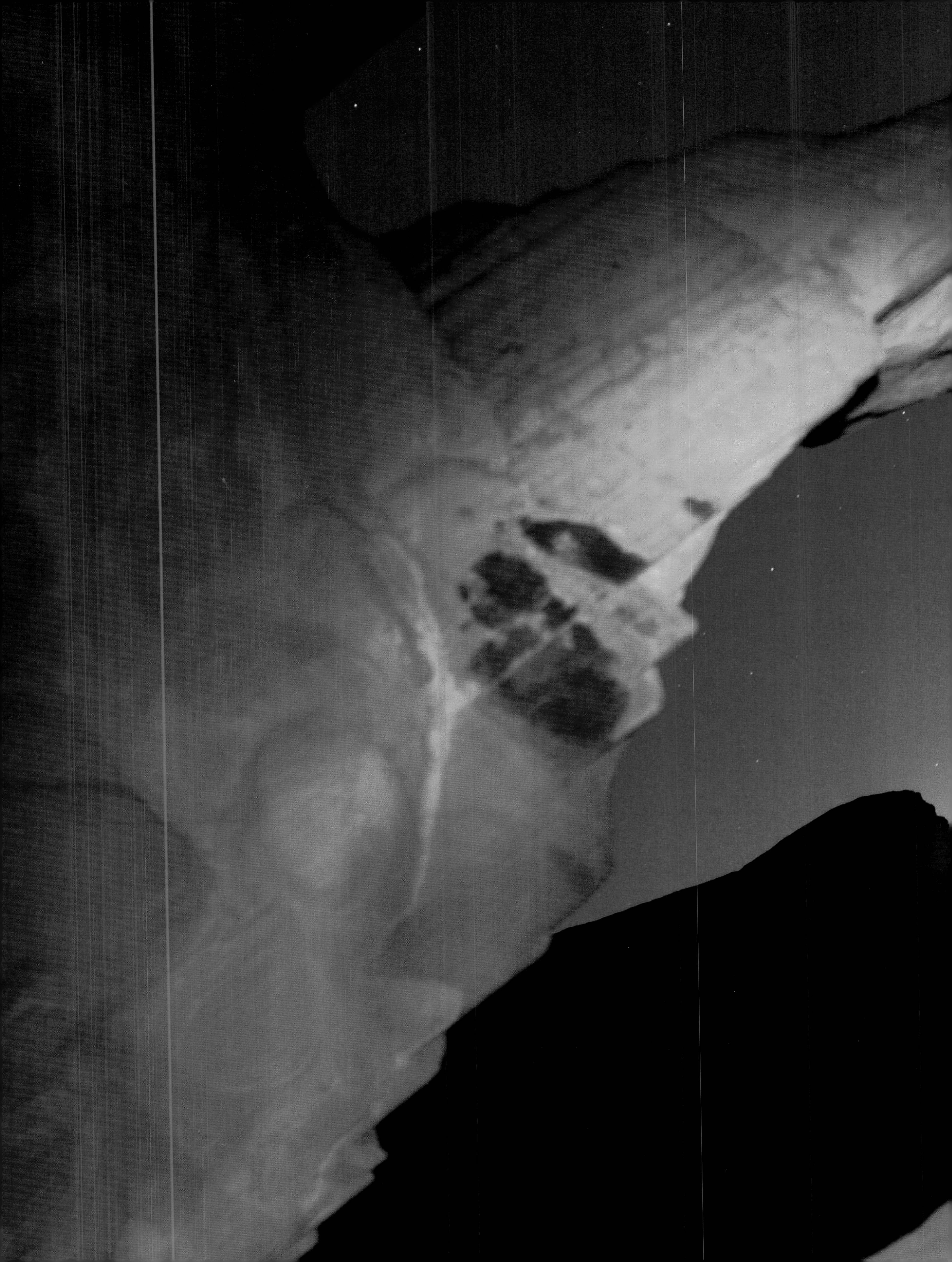

◁ Moonrise over the sea

THE GLOWING FULL MOON, here rising over the Mediterranean Sea, illuminates clouds in Earth's atmosphere above it, producing crepuscular rays—which stretch like endless fingers into the clear sky above. These rays, created by patterns of clouds in our atmosphere, were named after the Latin word for "twilight." The Moon looks disproportionately large. That is because you are seeing it low in the sky, against nearby objects—like mountains—and subconsciously comparing them in size.

Previous pages | Moonrise over land

ANOTHER ARCH FRAMES THIS VIEW OF THE MOON, photographed at moonrise. The rocks on Earth—as seen here in the valley of Fire State Park, Nevada—are very different from those on the Moon. The sandstone making up Arch Rock was laid down in warm seas on Earth many millions of years ago. In contrast, the Moon has very little water (possibly a little ice near its south pole), and its geology is a challenge to understand.

Earthrise Apollo 8

IN 1968, FROM JUST ABOVE THE SURFACE OF THE MOON, the astronauts of Apollo 8 captured this spectacular image of an earthrise. It emphasises that there could hardly be more contrast between these two worlds that are only 238,855 miles/384,400 km apart. While the Moon is dead and lacks any atmosphere, Earth—our Blue Planet—has oceans, clouds, and a life-giving atmosphere of oxygen. This image shows the continent of Africa hovering over the barren lunar plains.

Earthshine

EVOCATIVELY NAMED "THE OLD MOON IN THE NEW MOON'S ARMS," this image of the waxing crescent moon also dimly reveals the rest of the Moon's disk. While the Sun is directly illuminating the crescent, "earthshine"—reflected sunlight from our own planet—is lighting up the remainder of the Moon. It is a very common sight in twilight skies when the crescent is slender—especially if cloud cover on Earth is dense—and our planet is therefore highly reflective.

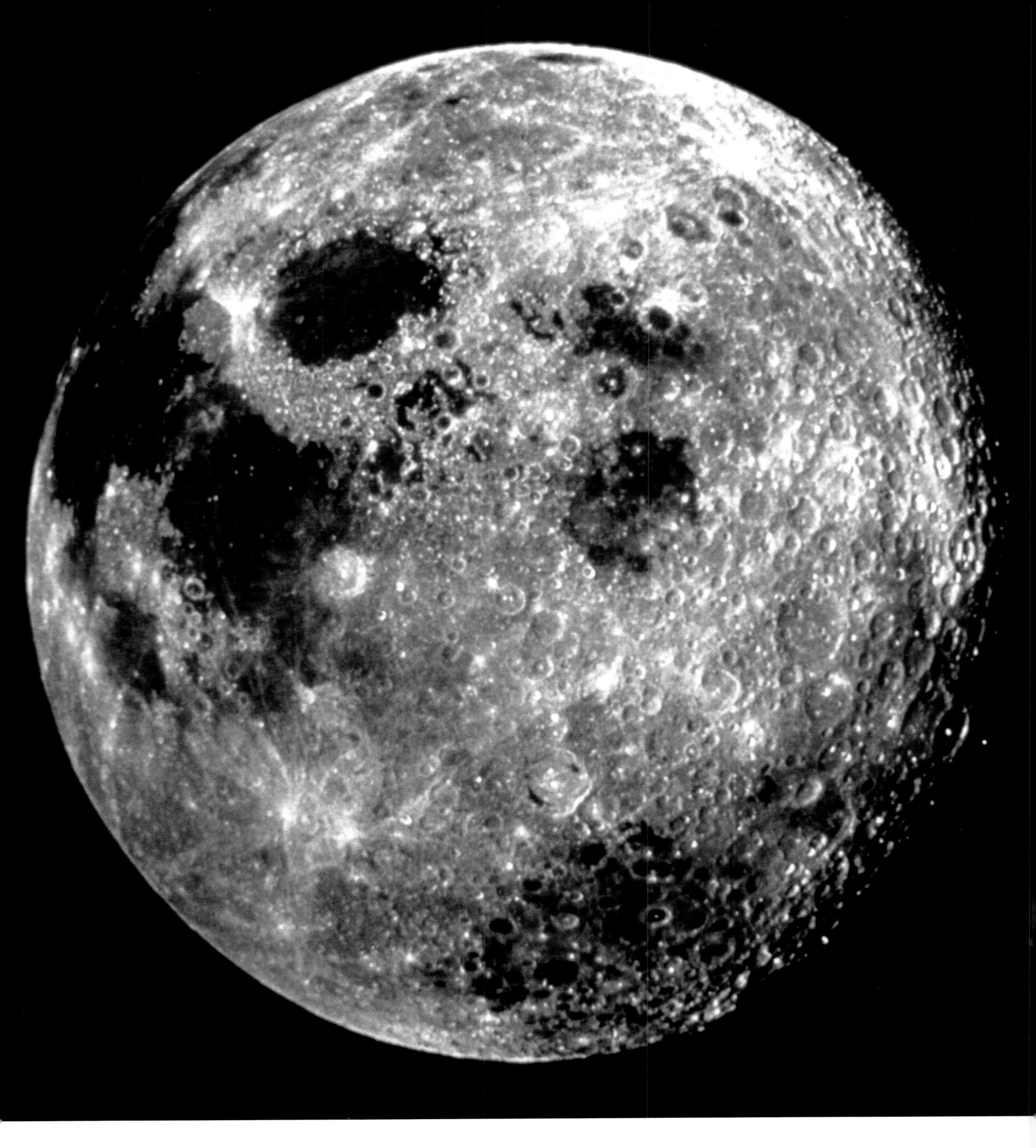

Full moon from space

IMAGE OF THE NEARLY FULL MOON AS SEEN FROM SPACE. At top left is Mare Tranquillitatis, where the first Apollo astronauts landed in 1969. The "seas" are huge impact basins. A Welsh astronomer, Sir William Lower, observed the Moon through a telescope even before Galileo did and described its surface as being "like a tart that my cook has made . . . here a vein of bright stuff, and there of dark, and so confusedly all over. I must confess I can see none of this without my cylinder."

Phases of the Moon

EVERY 29.5 DAYS, THE MOON CIRCLES THE EARTH. It has no light of its own but reflects the sunlight that falls on its surface. This sequence of photographs shows a complete lunar cycle. When the Moon is in line with the Sun, it is "new." Then it gradually moves around the Earth, becoming ever more illuminated ("waxing"), until it reaches full moon. The phases then go into reverse ("waning"), until the Moon reaches new again.

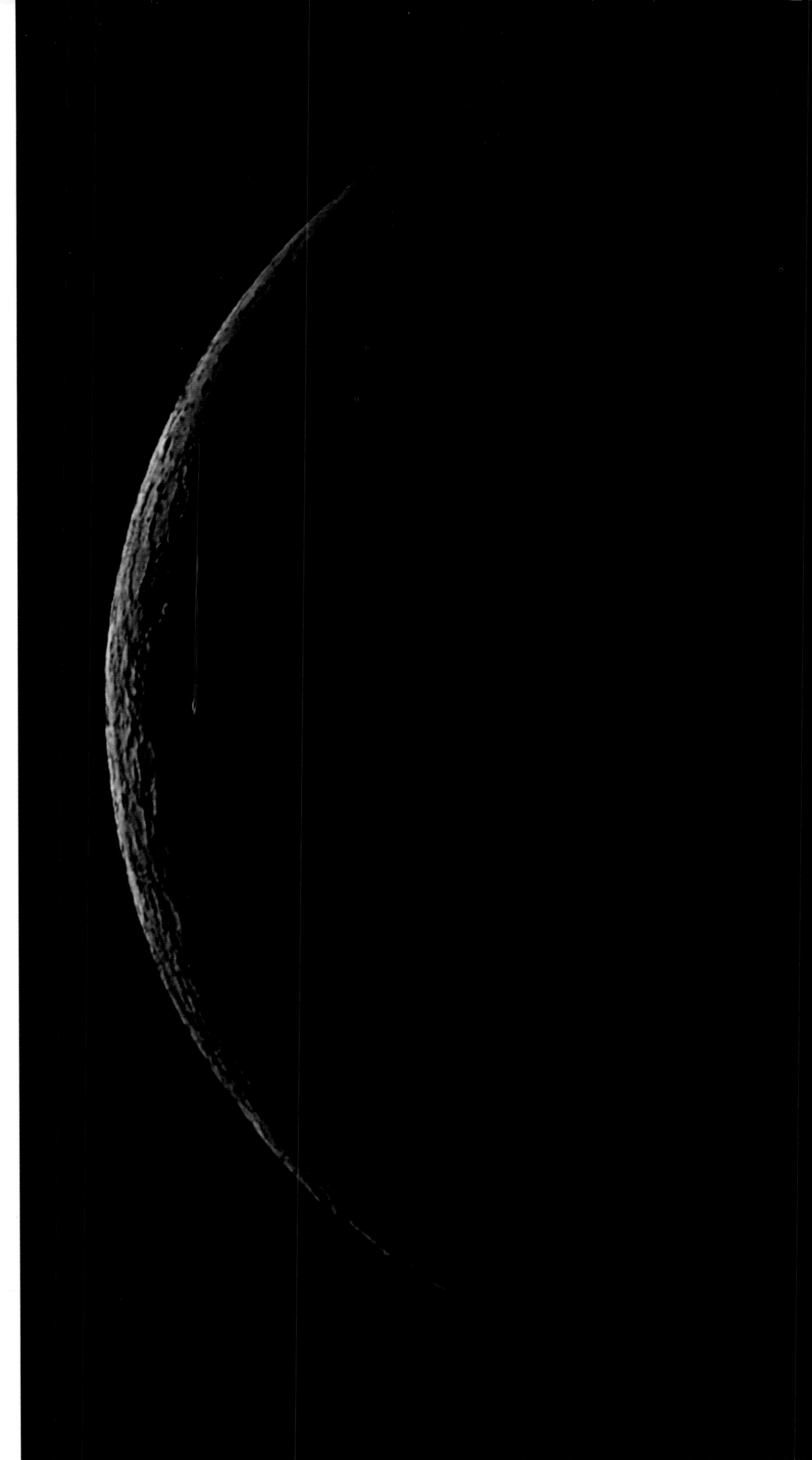

Waning crescent moon

WITH JUST ONE DAY TO GO BEFORE NEW MOON, this thin crescent was captured twenty-eight days into the lunar cycle in October 2001. When the Moon is illuminated from the side—rather than front-on, as at full moon—its dramatic features are far more apparent. Here, the shrinking limb of the Moon is pockmarked with craters, the legacy of countless impacts. The long shadows that the crater walls and mountains cast emphasize its rugged, airless landscape.

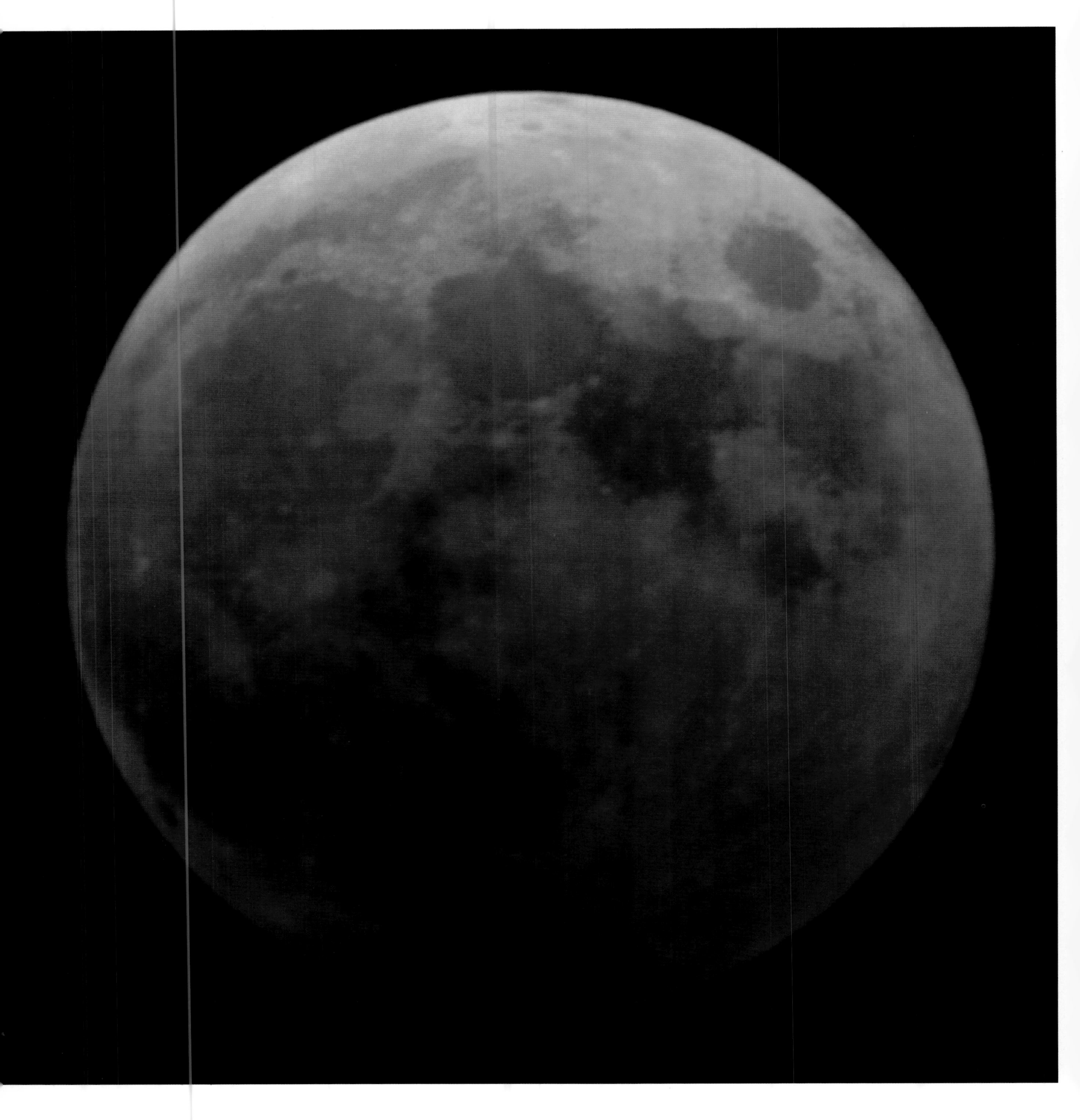

Lunar eclipse at totality

THE MOON'S ORBIT IS ANGLED TO THAT OF THE EARTH'S, so that when it goes behind our planet, the Sun's rays can usually still fall on it—and we see the Moon as full. But about twice a year, the Moon moves into the Earth's shadow and we see a total lunar eclipse. The Moon can become completely dark, but often sunlight is refracted around by Earth's atmosphere, especially if there is volcanic dust suspended there—giving the Moon's disk a baleful red color, as in this image.

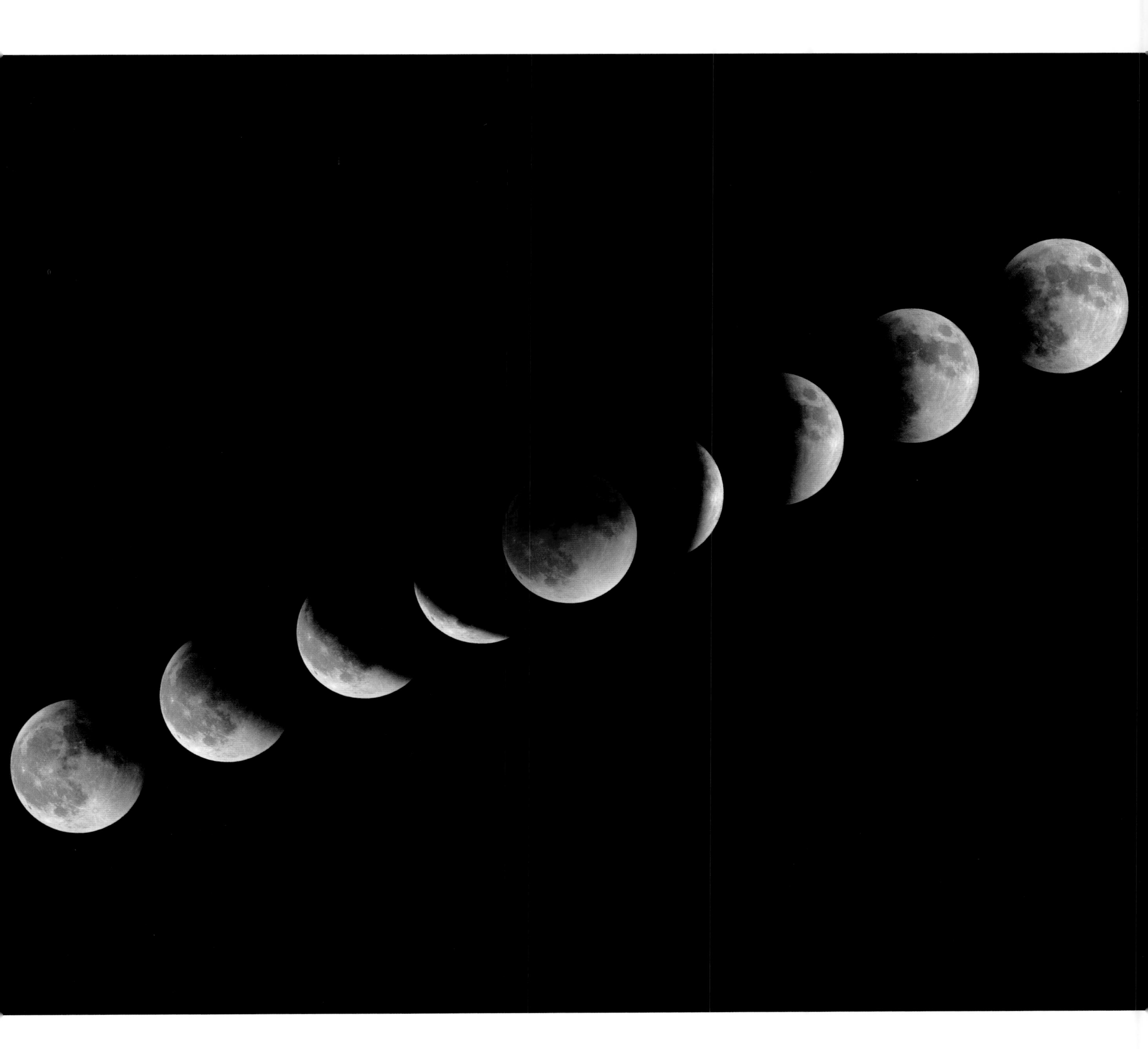

Lunar eclipse sequence

OBSERVERS IN DAYTON, OHIO, captured this sequence of time-lapse images of the total lunar eclipse on May 20, 2003. It graphically demonstrates how the Moon slides into the Earth's shadow before moving on again. (The central image of totality is overexposed.) The curvature of the shadow also reveals that the Earth is about four times the width of the Moon. Eclipses of the Sun and Moon have a profound effect on animals. During the hour-long totality of a lunar eclipse in Colombia, the local bullfrogs completely stopped croaking.

Moon crater Apollo 13

ALTHOUGH THE APOLLO 13 MISSION WAS SERIOUSLY THREATENED by the explosion of an oxygen tank on board—risking the astronauts' lives—it didn't stop the crew from taking photographs as they swung around the Moon. This image of a young crater near the northeast part of the Moon was taken as the astronauts sped past. The bright ejecta blanket of blasted-out material testifies to the crater's relative youth.

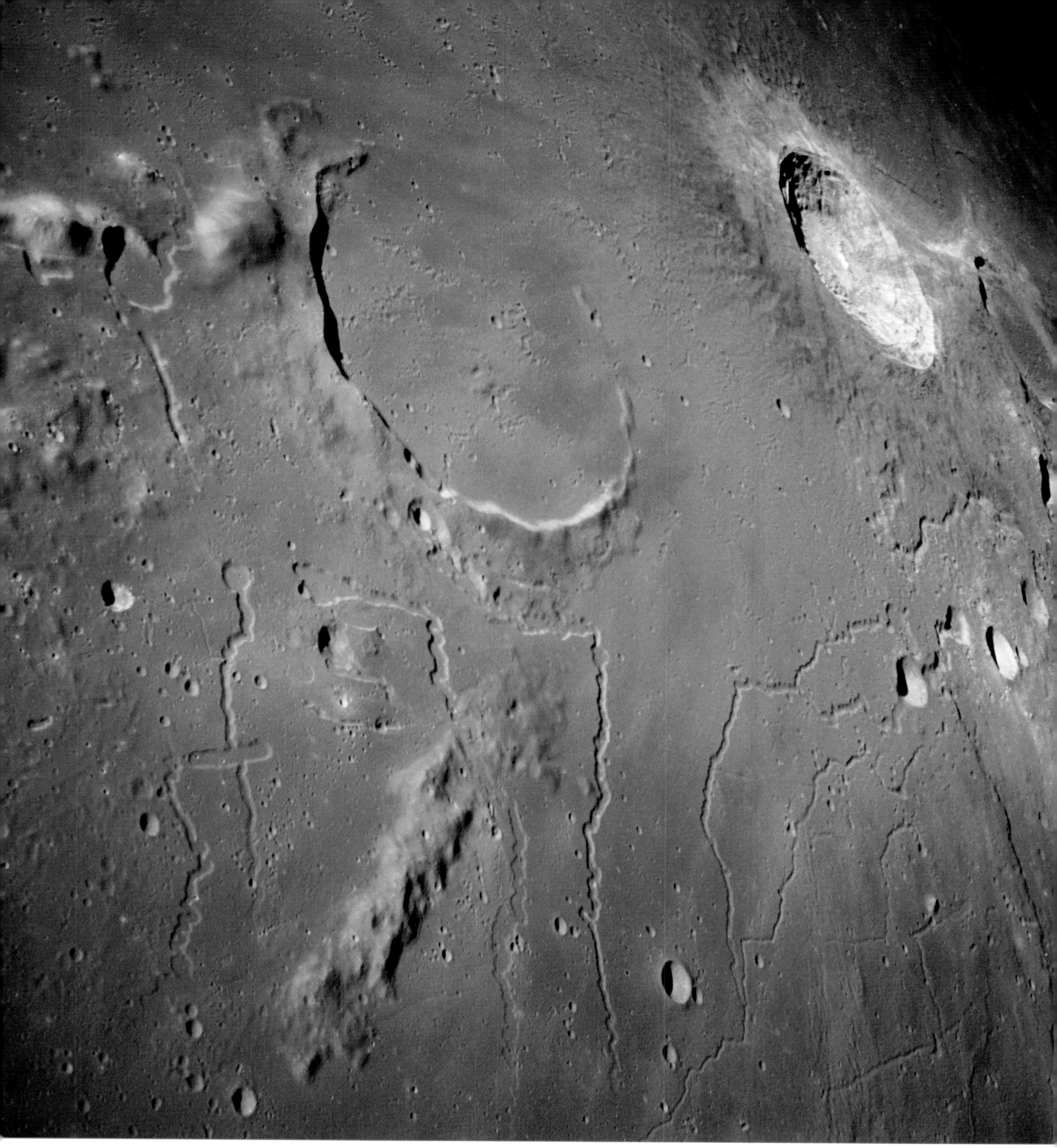

Lunar surface Apollo 15

THIS IMAGE, TAKEN BY THE APOLLO 15 ORBITER, shows the 6-mile- /42-km-diameter crater Aristarchus (top right) in the Oceanus Procellarum. It was formed less than a billion years ago and is surrounded by a halo of brilliant debris that enables it to be seen with the unaided eye. The sinuous channels (lower part of picture) are collapsed lava tubes, and the eroded crater (center) is very old.

Moon landscape Apollo 15

AT 15,100 FEET/4,600 METERS, MOUNT HADLEY dominates this image from the Apollo 15 mission in 1971. It is part of the Apennine mountain range, which forms the rim of the circular Mare Imbrium, blasted out by a giant asteroid 3.8 billion years ago. The astronauts used a vehicle a lot like a golf cart, the Lunar Reconnaissance Vehicle (LRV), to explore the surface on this mission, and its tracks are visible in the foreground.

Astronaut Apollo 16

ASTRONAUT BUZZ ALDRIN—THE SECOND MAN ON THE MOON—described its surface as "magnificent desolation." This is certainly true of the vista here, photographed on the Apollo 16 mission, in which Charlie Duke bores into the ground to take a vertical soil sample. Behind him is a small crater about 98 feet/30 m across. Duke and his colleague John Young (later to be the first commander of the Space Shuttle) spent seventy-one hours on the Moon and traveled 16 miles/26 km in the LRV. Apollo 16 was the penultimate crewed mission to the Moon, returning to Earth on April 27, 1972.

Camelot Crater Apollo 17

THE LAST HUMAN MISSION TO THE MOON—APOLLO 17—was the only expedition to have on board a qualified geologist, Harrison ("Jack") Schmitt. NASA targeted the Taurus-Littrow region, known for its diverse rocks of all types and ages. Behind the LRV is the 1,970-foot- /600-m-diameter Camelot Crater. Schmitt and his colleague Gene Cernan stayed on the Moon for seventy-five hours and collected 243 pounds/110 kg of moon rocks. Samples of moon rocks were sent out to universities and institutions all over the world and are still being analyzed today.

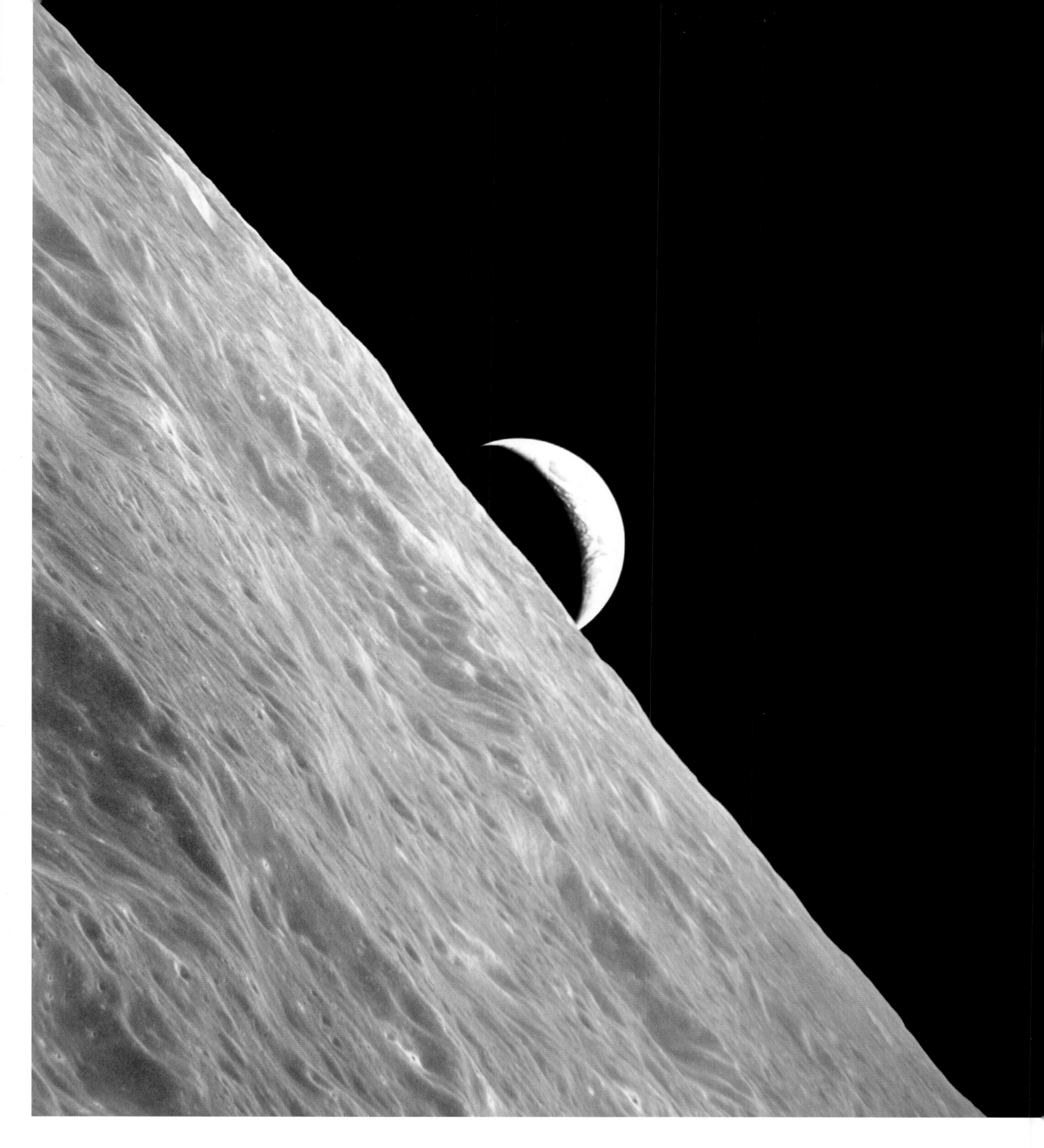

Earthrise Apollo 17

FAREWELL TO AN ERA: In December 1972, Ron Evans—Apollo 17's Command Module pilot in lunar orbit—captured this nostalgic portrait of the Earth rising over the far side of the Moon. Since then, no one has visited the barren plains of our sister world. Although robot probes have visited our companion in space, humans have not kangaroo-hopped or driven over its parched rocks for over thirty years, and Gene Cernan remains the last man to have stood on its bleak surface. But all this will change, as many nations unveil their plans to return humans to the Moon.

OUR BEAUTIFUL BLUE PLANET IS JUST ONE OF NINE WORLDS ORBITING OUR LOCAL STAR, THE SUN. It is unique in the solar system in that it has water and is teeming with life—but the other planets are no less individual.

Many people are amazed that most of the planets are brilliantly visible in the night sky. Yet our ancestors, thousands of years ago, were well aware of this. The ancient Greeks dubbed them *planetes*—meaning "wanderers"—because they saw them moving against the distant stars. For them this was confirmation that these worlds lay nearby.

The planets became revered as gods. Stately, slow-moving Jupiter was named after the king of the gods, blood-red Mars after the god of war, and diamond-hued Venus was celebrated as the goddess of beauty and love. In the image on these pages, we see a sensational shot of the two brightest planets—Venus and Jupiter—almost merging in the twilight sky.

Today, we respect the planets' ancient names, but we have a new understanding of their identity. Our neighbor-worlds fall into two groups: small rocky planets close to the Sun (Mercury, Venus, Earth, Mars), and huge, gaseous worlds further out (Jupiter, Saturn, Uranus, and Neptune). Distant Pluto—discovered as recently as 1930—remains an enigma. It may not be a stand-alone planet at all but part of a swarm of small bodies, like those in the asteroid belt, which lies between Mars and Jupiter.

The character of the planets moving outward from the Sun speaks volumes as to how they were formed. We know that the Sun was born in a nebula—a cloud of gas and dusty space-soot—some 4.6 billion years ago. Close to the hot, evolving Sun, only the dust could survive, where it consolidated itself to create the rocky innermost worlds of our solar system. But farther out, in the bitter cold of space, the fledgling planets wrapped themselves in blankets of gas from the nebula—forming the gas giants.

Until the middle of the twentieth century, the planets were just dots in the sky. Then the spacecraft revolution erupted, turning the dots into real worlds. Since then, the solar system has been saturated with a plethora of probes. Mars has been the focus of much of this exploration centered on the search for life.

Life may also exist on the myriad moons of our neighbors' worlds. Jupiter's Europa may harbor aquatic life under a deep ocean, while Titan—circling Saturn—could support life in the future, when the Sun grows warmer.

How typical is our solar system—and what are the chances for life beyond? In the past ten years, astronomers have discovered over 160 planets circling other stars. Because the techniques for detection are in their infancy and we can pick out only the biggest planets—which are unlikely to harbor life—it is too early to say. But new space missions scheduled for the future are being designed to seek out twins of Earth in these planetary systems.

The odds are that we inhabit a vibrant, living universe.

3 PLANETS

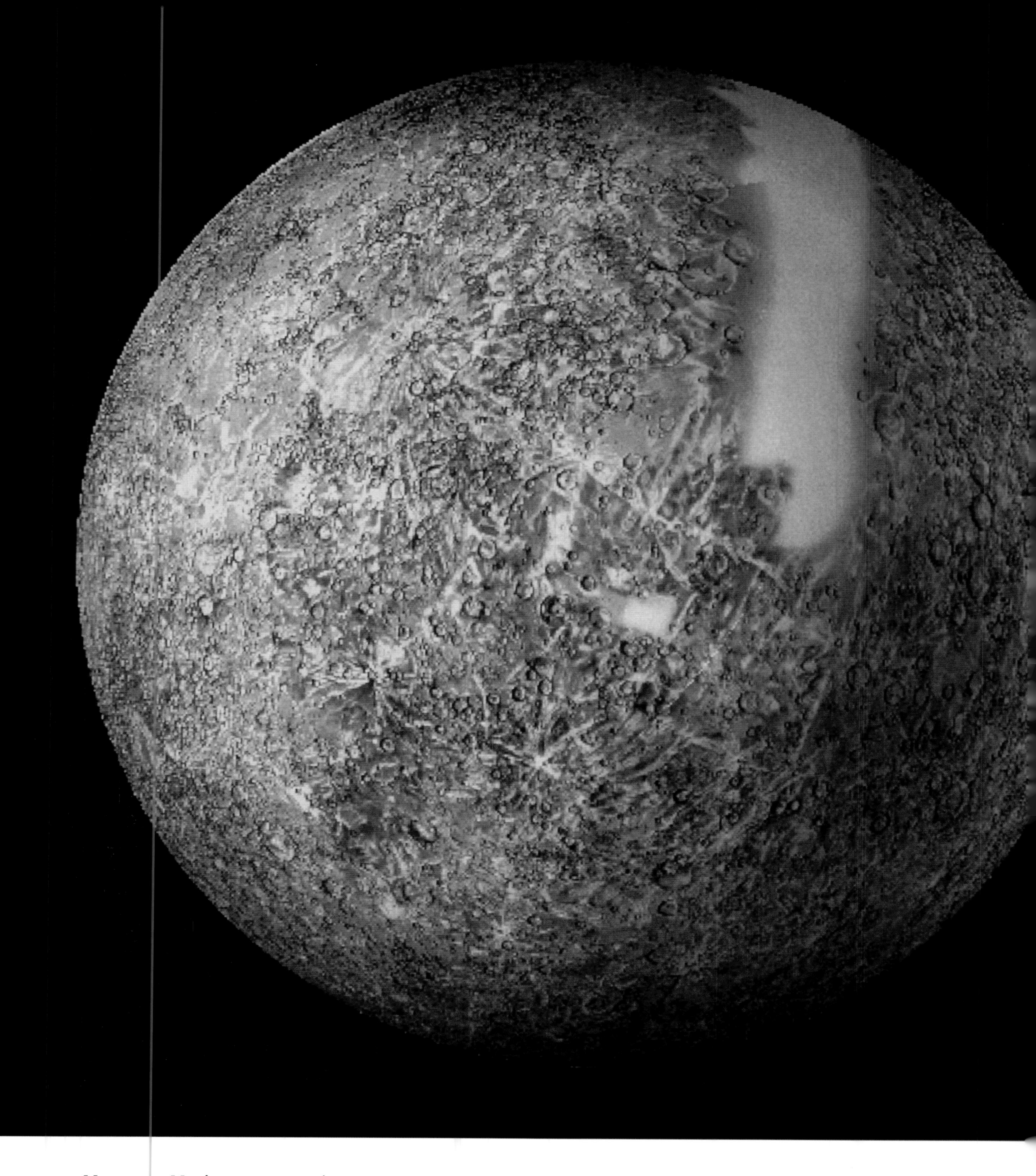

Mercury: Mariner 10 mosaic

THE ONLY SPACE PROBE TO VISIT MERCURY, the Sun's innermost planet, was Mariner 10—which swung past in 1974 and 1975. It revealed a cratered world very like our Moon, both in size and appearance. But Mercury has a dense iron core, and researchers believe that the cooling and shrinking of this core may have led to the formation of enormous "wrinkle ridges" on the planet (center)—like the creases on a dried-up apple. The blank portions of this image are regions where the probe was unable to obtain data.

Previous pages | Our planetary neighbors

JUPITER AND VENUS GET UP CLOSE AND PERSONAL TO ONE ANOTHER in the twilight skies of November 2004. Viewed from California, Venus—the brighter of the two worlds—is our nearest planetary neighbor, while mighty Jupiter (above) is the biggest planet of all. Planets shine only by reflected sunlight, but they are among the brightest objects in the sky. Our neighbor-worlds move slowly and majestically against the background of stars, they and have inspired many cultures to regard them as deities.

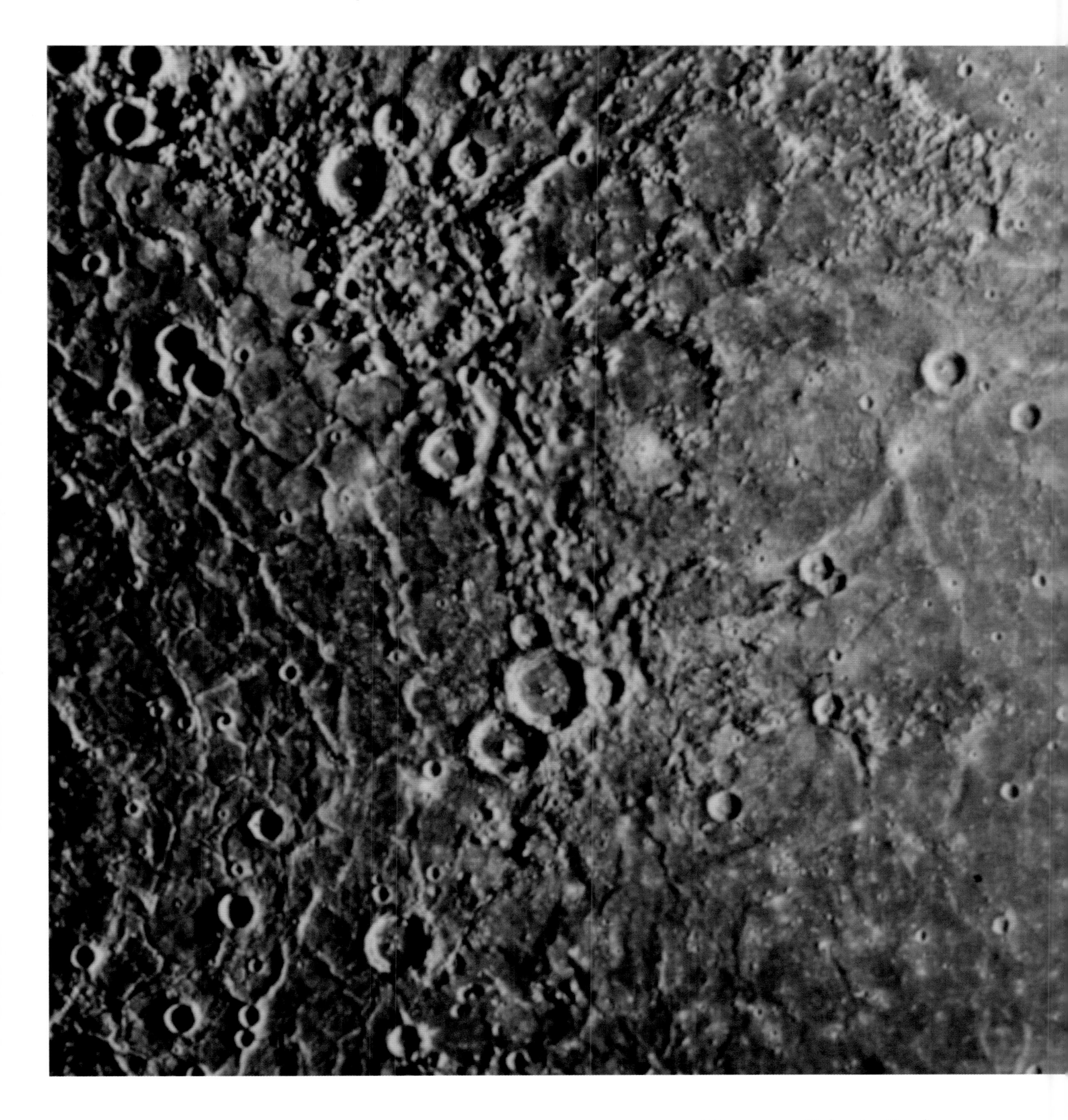

Mercury: Caloris Basin

DENSELY CRATERED MERCURY HAS BEEN BATTERED BY METEOROIDS since it was born. But the ultimate impact was caused by a 62-mile/100-km asteroid, which formed the Caloris Basin (left of image). The enormous multiringed structure is 808 miles/1,300 km across, and it is the biggest feature on this small planet. It derives its name from its situation at the subsolar point (where the Sun's rays hit Mercury most directly). Temperatures here can reach 806 degrees Fahrenheit/430 degrees Celsius.

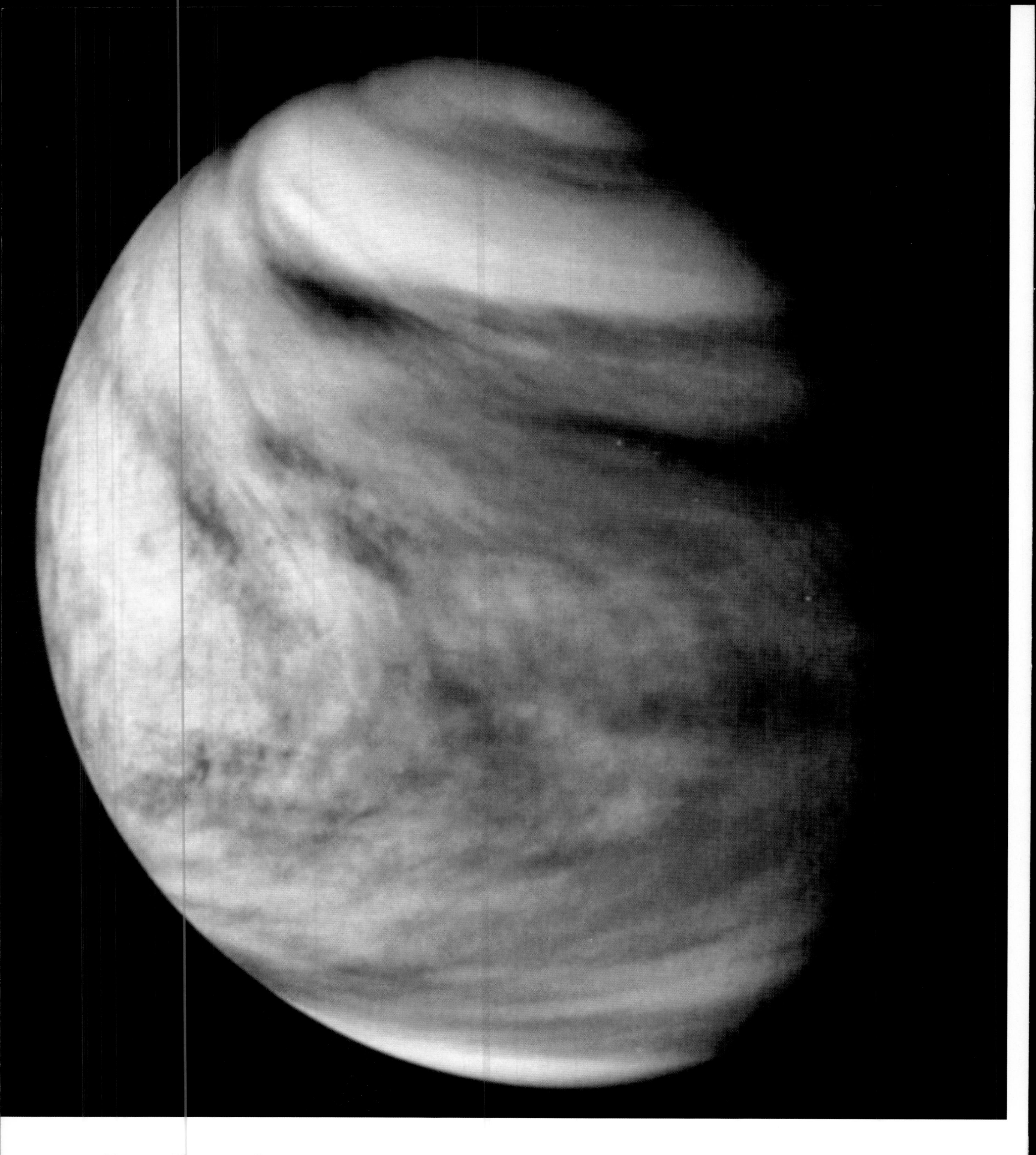

Venus: Pioneer photo

VENUS—THE PLANET THAT CELEBRATES THE GODDESS OF BEAUTY AND LOVE—is anything but. This Pioneer Venus Orbiter image reveals a thick carbon dioxide atmosphere, laced with clouds of sulfuric acid. The atmosphere is so heavy that the pressure at Venus's surface is ninety times that of Earth's. And the carbon dioxide in the atmosphere has led to a runaway greenhouse effect, making it the hottest planet in the solar system (869 degrees Fahrenheit/465 degrees Celsius). If you visited Venus, you would be roasted, crushed, corroded, and suffocated.

Moon and Venus ▷

DAWN: THE MOON AND VENUS RISE OVER A LAKE. There is no disputing that Venus is a glorious sight in the sky. Because of its reflective cloud layers—and its proximity to Earth (it is our nearest planetary neighbor)—it can look like a small lantern at sunrise and sunset. Venus is by far and away the brightest planet, and it can even cast shadows. Because it lies closer to the Sun than the Earth, it never strays far from our star in the sky. That is why Venus is often called "the morning star" or "the evening star."

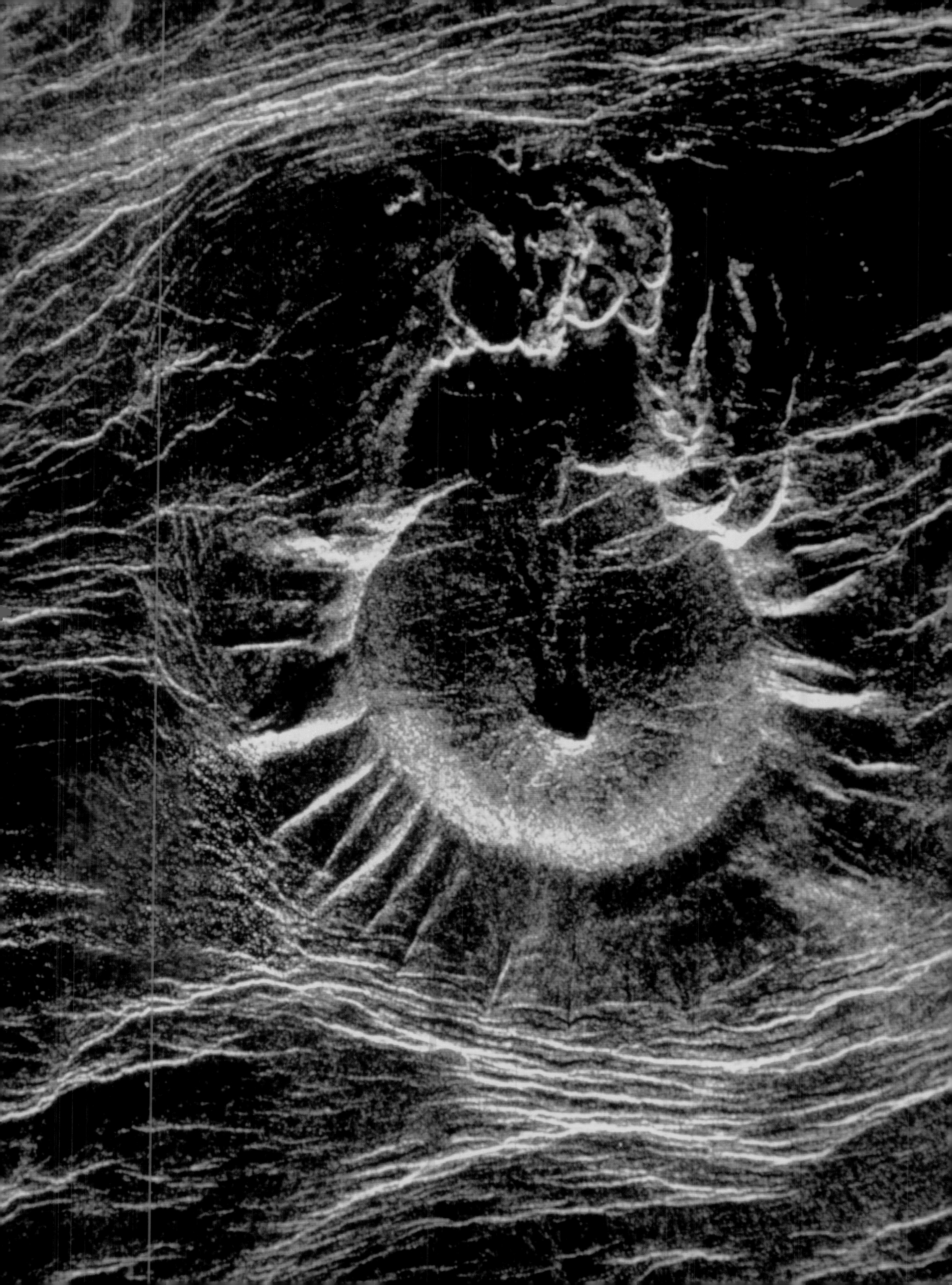

◁ Venus: volcano in Estia Regio

VENUS IS COVERED IN VOLCANOES, like this one in the Estia Regio region in the southern hemisphere of the planet. The volcano is 41 miles/66 km wide, and its crater is 22 miles/35 km in diameter. At the top, a lava flow tumbles down from the summit. Researchers do not know if the volcanoes on Venus are currently active—although they strongly suspect that they are—because this would account for the sulfuric acid levels in the atmosphere. This false-color radar image was captured by NASA's orbiting Magellan probe in 1991.

Venus: radar map

MAGELLAN'S RADAR PENETRATED THE DENSE CLOUDS OF VENUS to map its surface. In this false-color view, the "continents" of our neighbor-world emerge. Blue regions are areas below the "mean" radius (rather like oceans on Earth); the colored regions are higher. The large landmass at the top is an upland "continent" called Ishtar Terra. Scientists are still unclear as to whether Venus's landforms are static, or whether they are driven by internal currents called plate tectonics (as on Earth) and pushed around the globe.

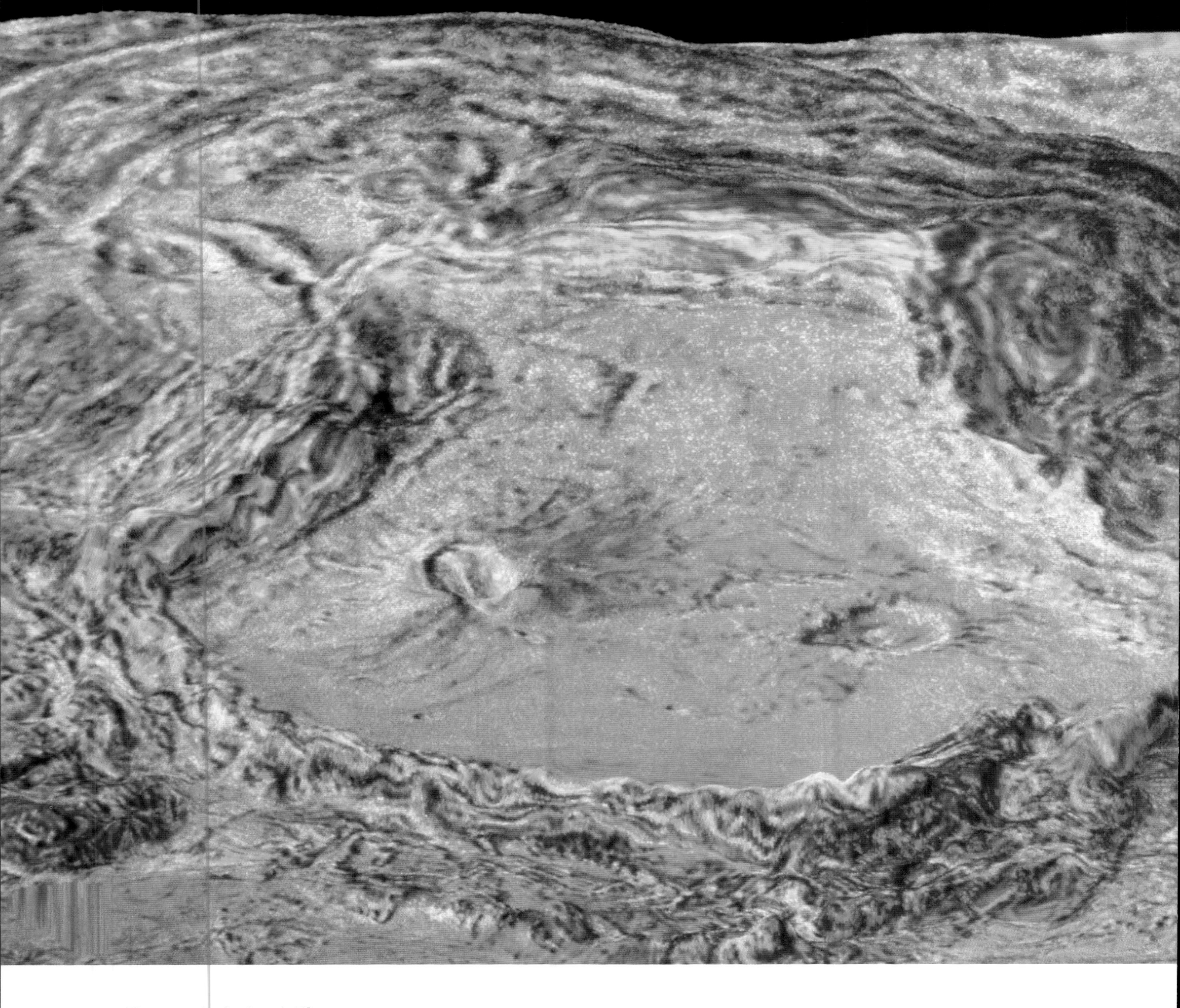

Venus: Lakshmi Planum

THIS FALSE-COLOR IMAGE HOMES IN ON LAKSHMI PLANUM, part of the "continent" of Ishtar Terra. At the center are two large volcanic calderas: Sacajawea (right) and Colette (left). The Akna Mountains (orange) lie above the calderas, while Maxwell Montes (also orange) is at the far right. Towering 7 miles/11 km above the plains, Maxwell—named after the Scottish physicist James Clerk Maxwell—is the highest point on Venus. It is also the only feature on the planet to have been named after a man instead of a woman or a goddess.

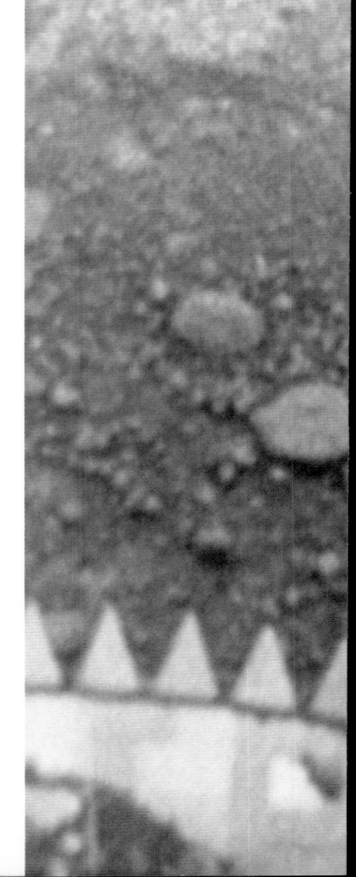

Venus surface

SEVERAL RUSSIAN VENERA CRAFT HAVE BATTLED THE HIGH TEMPERATURES and pressures of Venus to make it to the surface—where they have survived, albeit briefly. In March 1982, Venera 13 landed in a region called Beta Regio, amid a barren scene of flat volcanic rocks. This image also reveals part of the landing craft as well as the dull orange glow that pervades the planet. The probe lasted for two hours and seven minutes—capturing fourteen images—before succumbing to the atmospheric pressure.

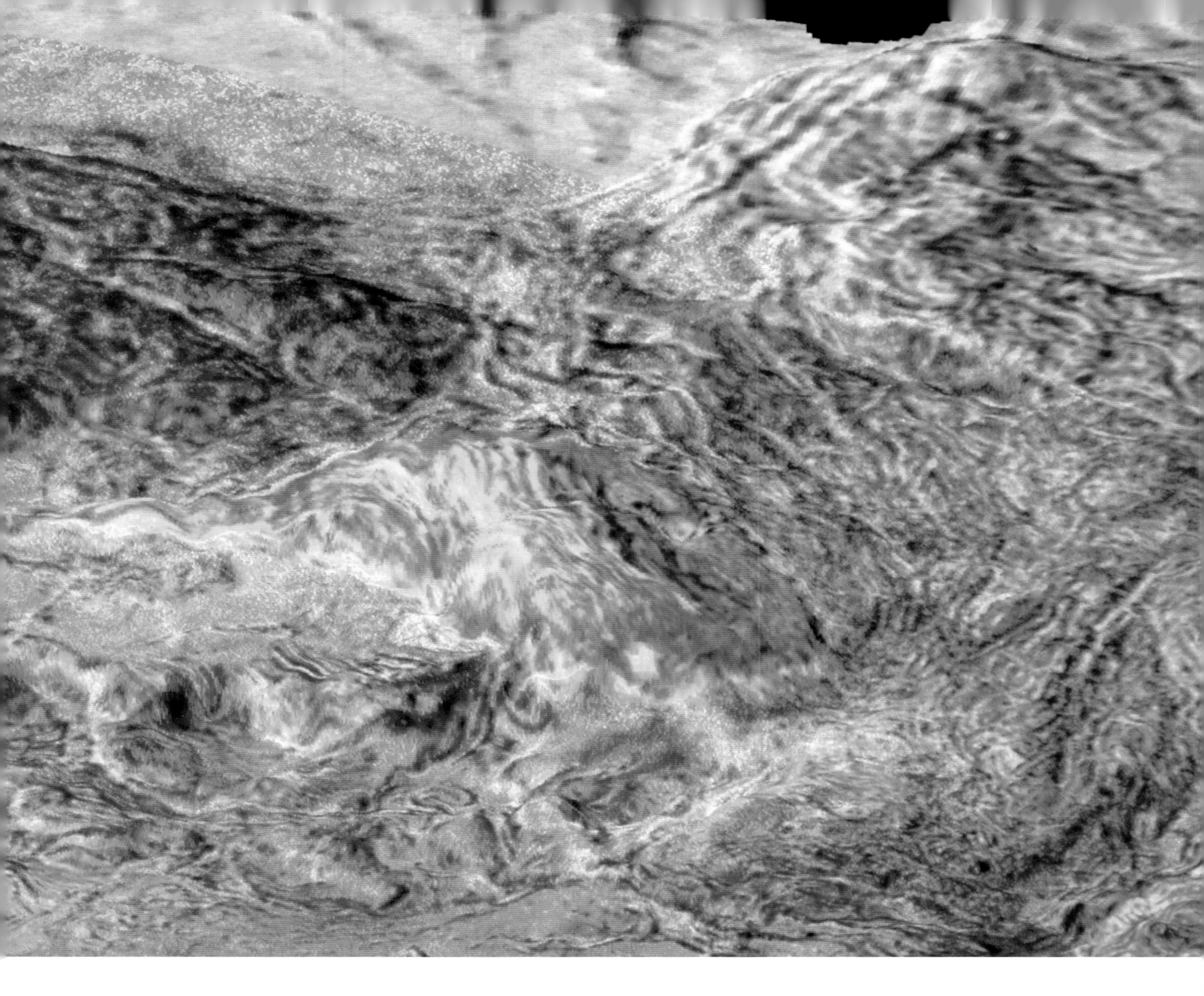

Whole Earth Apollo 17

THIS CLASSIC PHOTOGRAPH OF THE EARTH—taken by Gene Cernan on the final human Apollo mission to the Moon in 1972—may well have helped to inspire the whole green movement. Our planet looks like a fragile blue ball, shielded by a precariously thin atmosphere, which floats in the hostile blackness of space. The image reveals the continent of Africa and the Antarctic ice cap. But even more, it reveals the vulnerability of our world and the huge diversity of life-forms that live on it.

Crescent Earth

GOODBYE, EARTH. The astronauts on the first manned lunar mission in 1969 captured this beautiful image of our crescent world. How did they feel as they looked down on our fertile planet, destined as they were to land on a barren, airless, unknown world—with possibly no chance of returning? Astronauts report that views of Earth from space are breathtaking. Hopefully—with the predicted rise of space tourism in the twenty-first century—more of us will be able to share the experience.

Mars: the Red Planet

MARS, THE NEXT PLANET OUT FROM THE SUN AFTER EARTH, is literally bright red. It is no wonder that our ancestors named it after the bloody god of war. Mars is, quite literally, rusty: the iron in its soil has reacted with water—which is almost certainly present today—and turned its rocks red. Although only just over half the size of Earth, Mars has more extreme geology than any other planet in the solar system. It is also the most Earth-like world orbiting the Sun—and it may well harbor primitive life.

Mars: Olympus Mons

THREE TIMES HIGHER THAN MOUNT EVEREST and large enough to cover Spain, Olympus Mons is the biggest volcano in the solar system. This 16-mile- /26-km-high colossus is part of a complex of gigantic volcanoes that were thought to be extinct but may just be dormant. The volcano is topped by an enormous caldera (collapsed crater), while cliffs thousands of feet high form its rim. Thin clouds in the sparse Martian atmosphere—largely composed of carbon dioxide—frame the view.

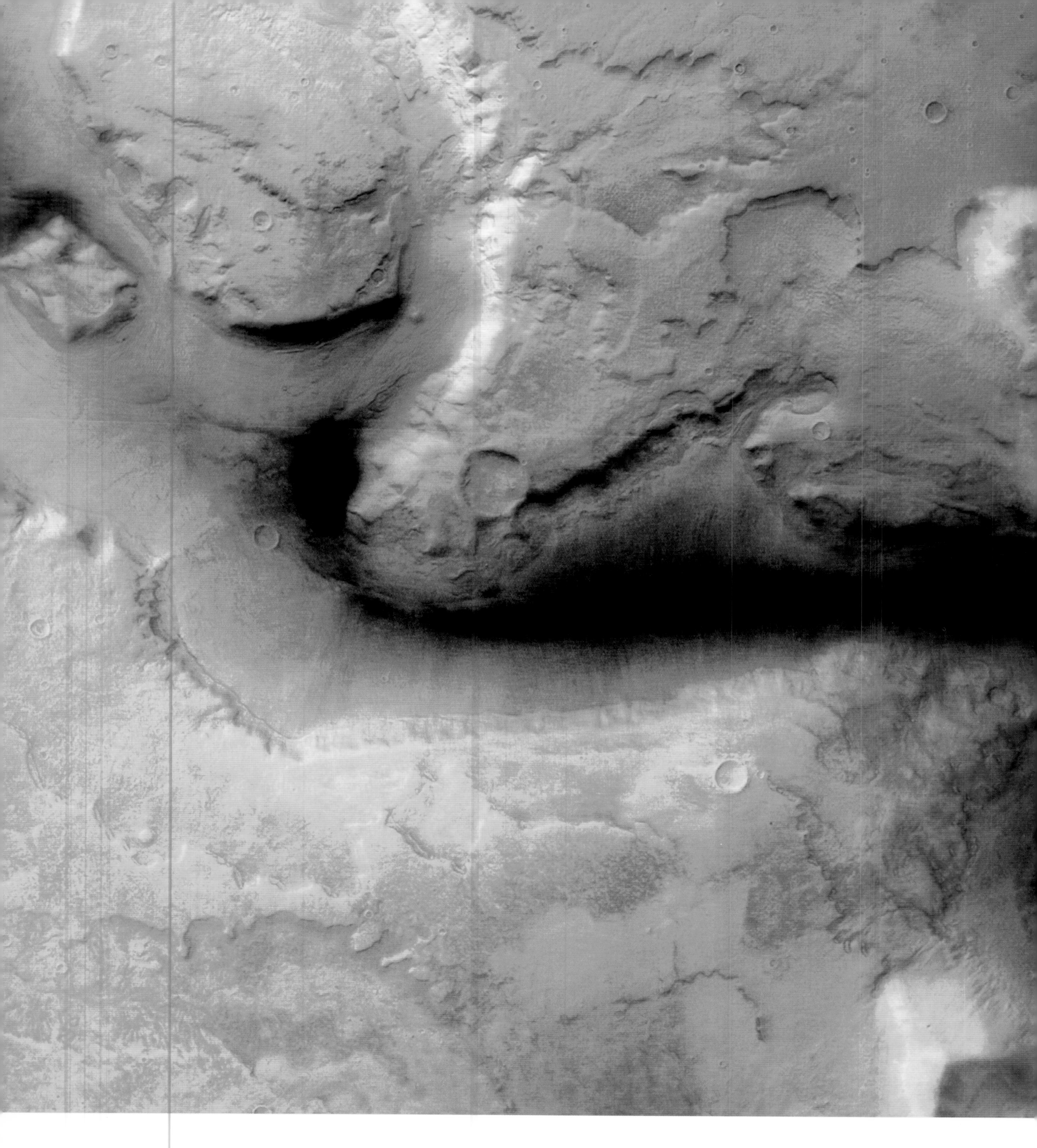

Martian surface

FROM A HEIGHT OF 170 MILES/273 KM, Europe's orbiting Mars Express probe photographed this image of a former water channel on Mars, covering an area around 62 miles/100 km across. Ruell Vallis is one of many thousands of sinuous watercourses that worm their way across the surface of the Red Planet. Water probably exists below the soil even today in the form of ice, and several future space-probe missions are designed to search for it. If we are to find life on Mars, water is a prerequisite.

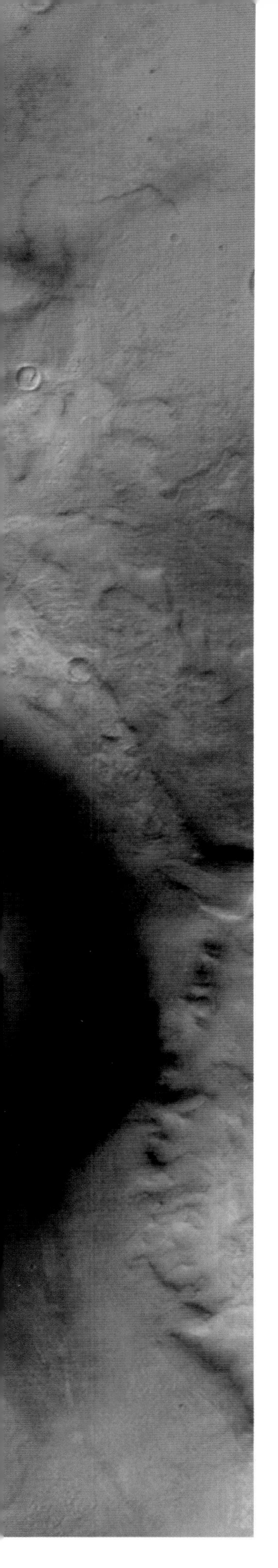

Mars: Adirondack

IN JANUARY 2004, EARLY ON IN ITS MISSION to explore the Red Planet's surface, NASA's Mars roving craft Spirit homed in on this large rock—nicknamed Adirondack. Scientists chose the rock because its clean, flat surface is ideal for analysis. Spirit has a grinding arm that can take rock samples and examine their mineral composition. Spirit's twin rover, Opportunity—which landed at a different site—carries the same equipment. Together, they have built up an astonishingly detailed overview of Mars's geology.

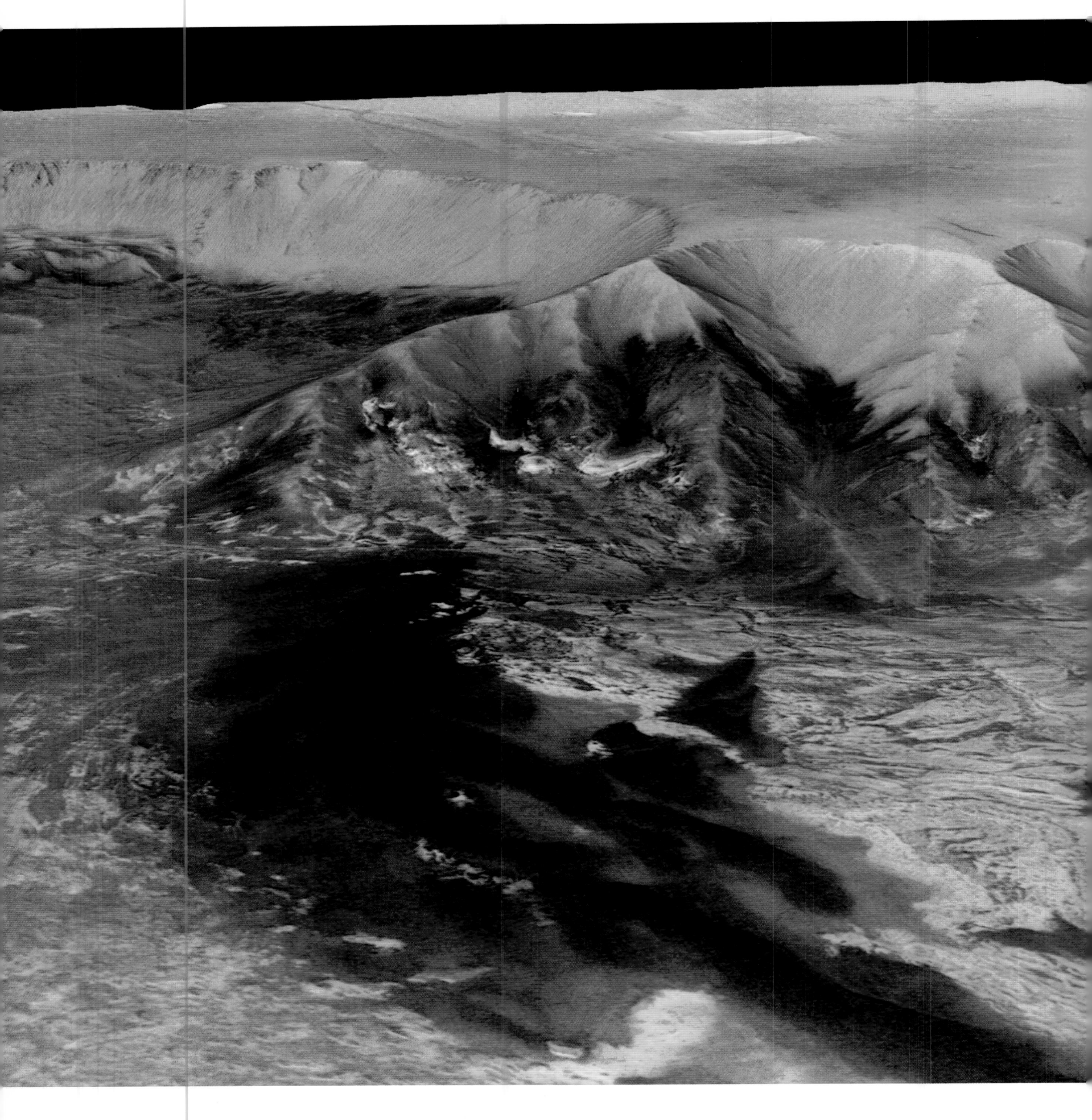

Mars: canyon rim

IN THE PAST, THE HUGE VOLCANOES ON MARS—which stretched and distorted its surface with their upwellling lava—literally split the world apart. As a result, the Red Planet is disfigured by a huge scar along its equator. Valles Marineris is over 2,486 miles/4,000 km long, ten times longer than the Grand Canyon in Arizona. This Mars Express image of part of the canyon (Melas Chasma) reveals a dramatic and forbidding terrain. In parts of the Valles Marineris, the cliffs are 4 miles/7 km high, and the branching canyons 124 miles/200 km wide.

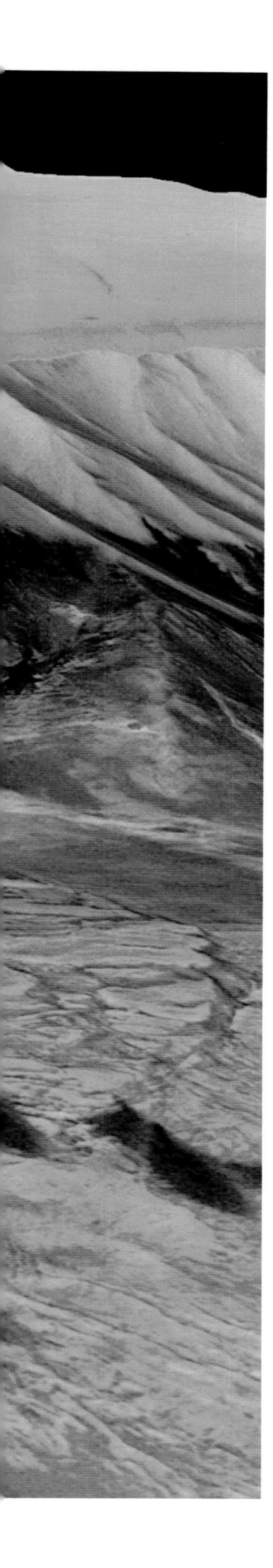

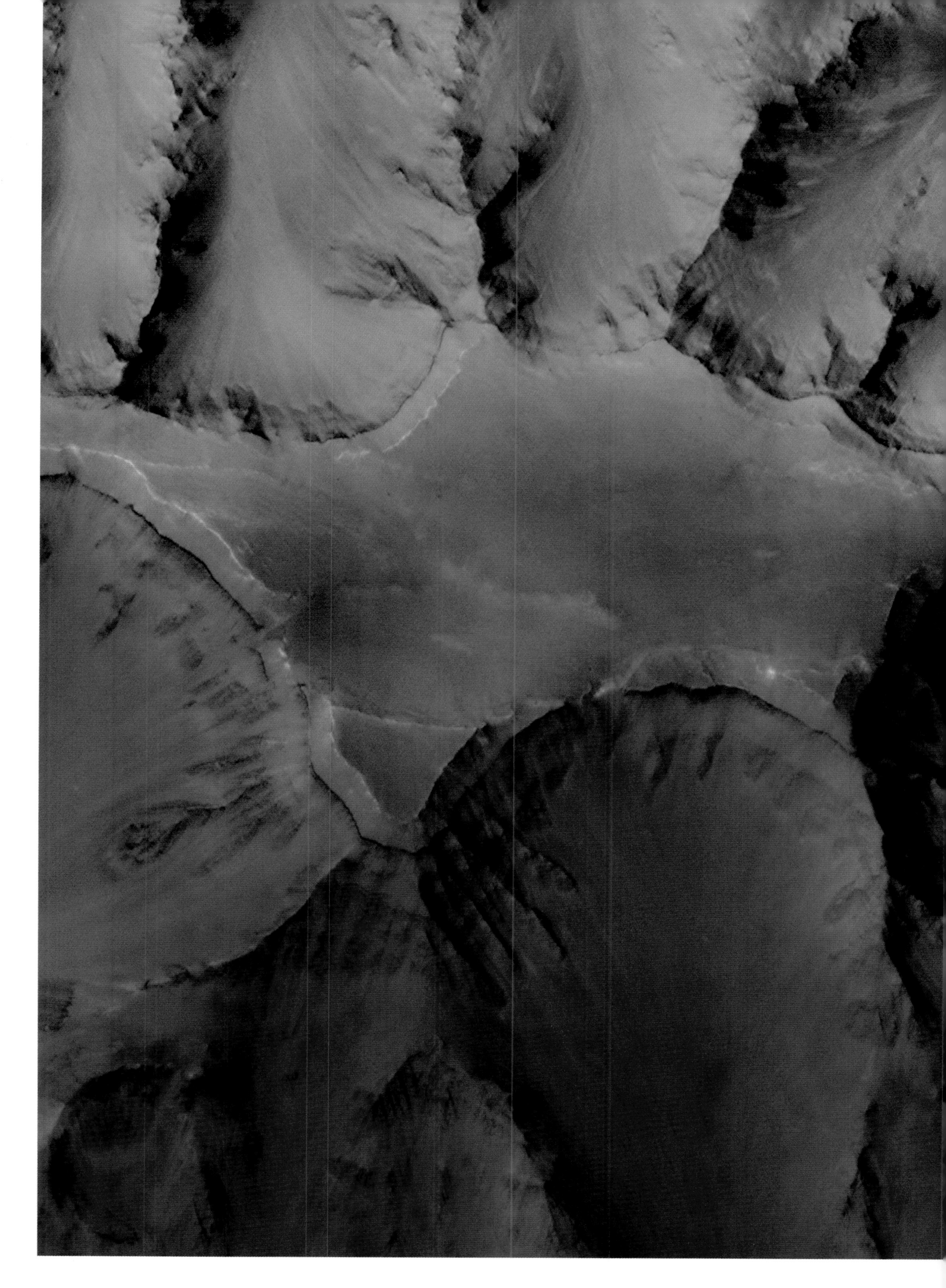

Mars: Valles Marineris

A STUNNING VIEW OF THE DEPTHS OF THE VALLES MARINERIS, captured by the Mars Global Surveyer probe. This spacecraft—which has been in orbit around the Red Planet since 1997—is known as the "stealth mission," because it has been largely unpublicized. It has taken some of the most detailed images of Mars ever, and continues to do so. This image reveals soft and eroded foothills surrounding the canyon. The terrain is layered, suggesting that water may once have flowed there.

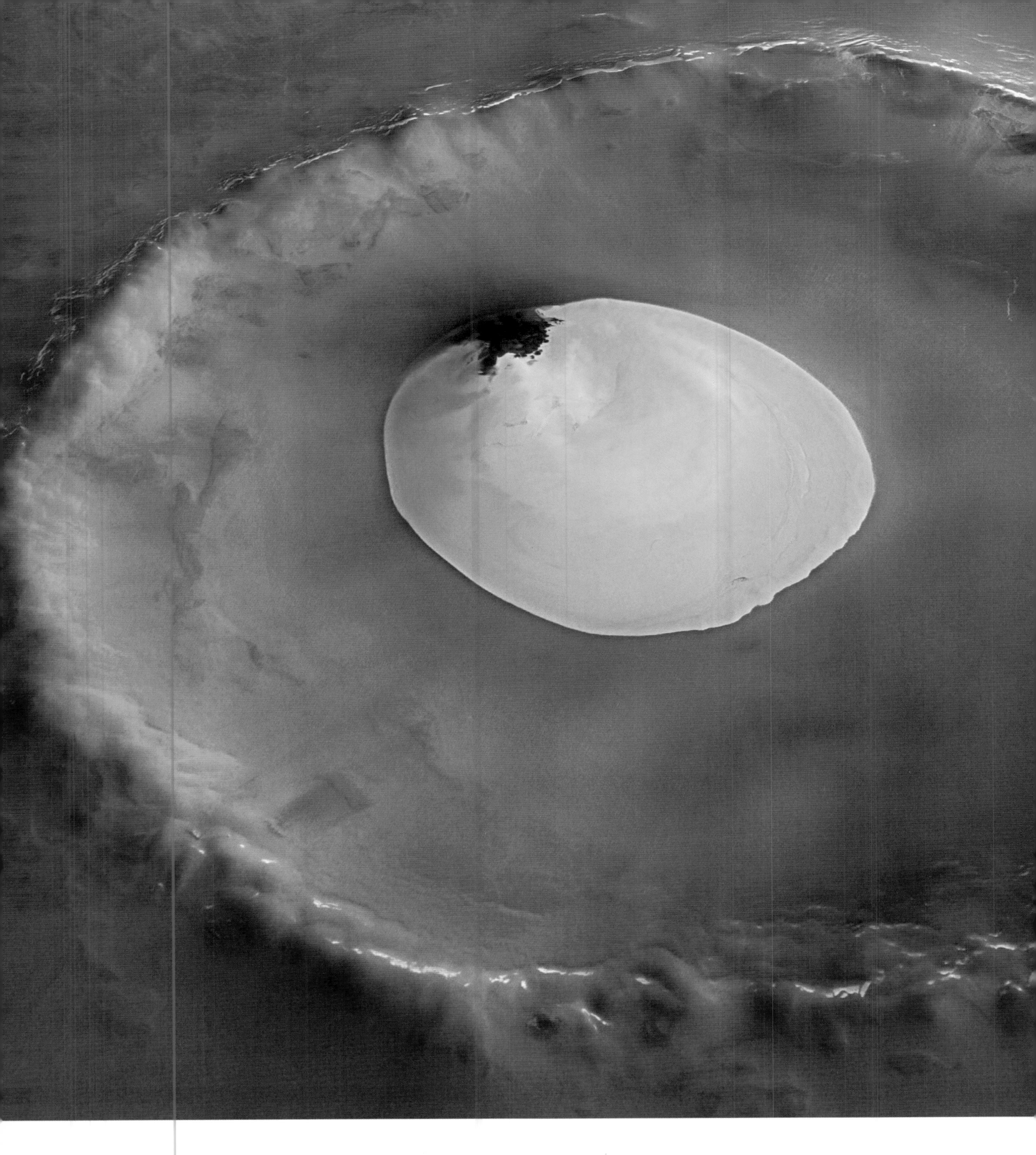

Mars: water ice in crater

A POOL OF ICE FORMS A VAST PUDDLE ON THE FLOOR of the crater Vastitas Borealis. At 22 miles/ 35 km across, this impact crater lies on a northern plain, where—as in similar locations on Earth—temperatures are low all year (even the average temperature on Mars is below zero). In the polar regions of the Red Planet, ice and frozen carbon dioxide linger all year. This true-color image was taken by the Mars Express Orbiter in February 2005.

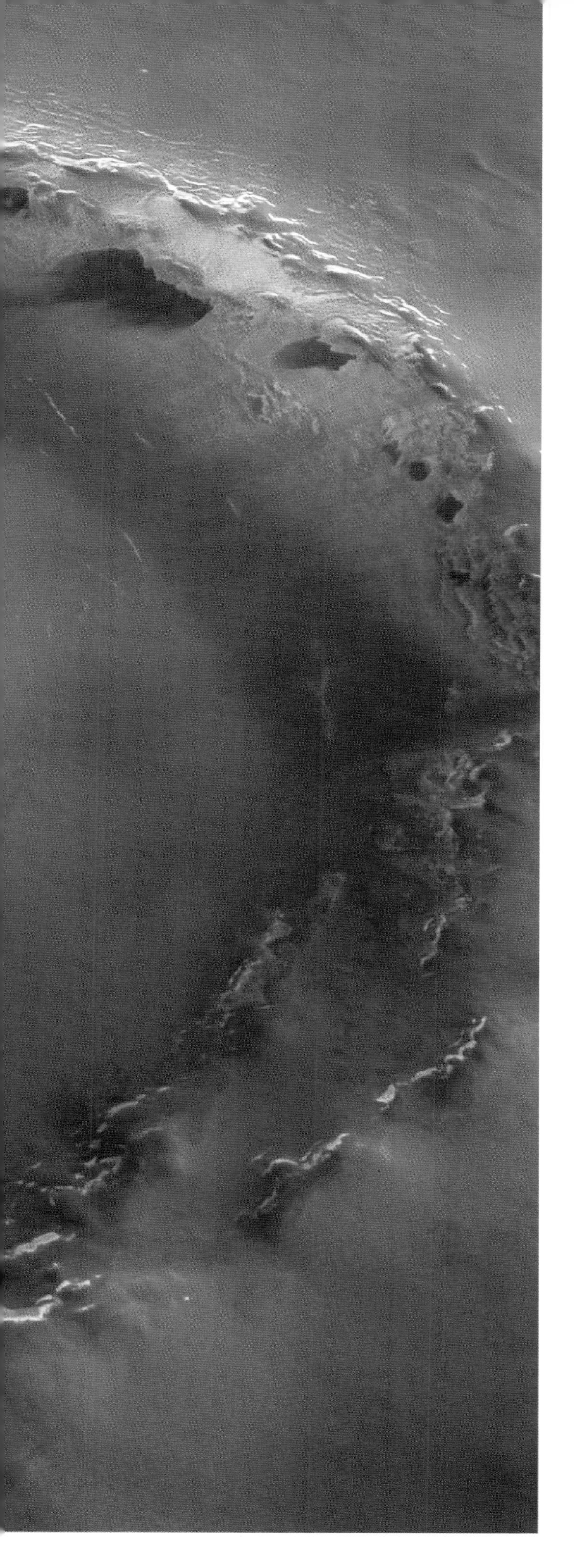

Martian moon: Phobos

MARS HAS TWO TINY MOONS—Phobos (meaning "fear") and Deimos ("panic"). Phobos, pictured here, is the larger of the two moons, yet it is less than 19 miles/30 km across. In this Mars Express image, Phobos is dominated by a 6-mile- /10-km-wide crater called Stickney. Just 5,841 miles/ 9,400 km above Mars, this moon orbits at breakneck speed, circling the Red Planet in just seven hours and forty minutes. It is gradually being pulled inward by Mars's gravity, and it will collide with the planet in about 50 million years' time.

Jupiter: Hubble image

ENORMOUS JUPITER IS NOTHING LIKE THE FOUR ROCKY INNER PLANETS. It is a vast mass of gas—although it may have a small solid core. It is so big that 1,300 Earths could fit inside it. Yet, despite its girth, it spins faster than any other planet in the solar system: its "day" is a mere nine hours and fifty-five minutes long. This rapid rate of rotation forces its equator to bulge outward, making Jupiter tangerine shaped, and stretches its cloud patterns into stripy belts of alternating colors. This image was captured by the Hubble Space Telescope.

Jupiter and moon

THE CASSINI SPACECRAFT IMAGED JUPITER and its biggest moon, Ganymede—one of at least sixty-three moons circling the Giant Planet—as it swung past on its way to Saturn. Jupiter's atmosphere is made up almost entirely of hydrogen and helium, with traces of methane, ammonia, and water. The higher white regions are called "zones"—the color comes from frozen ammonia crystals—while the lower-lying dark ocher areas are known as "belts." Through a small telescope, you can watch the belts and zones drifting across the planet as Jupiter spins.

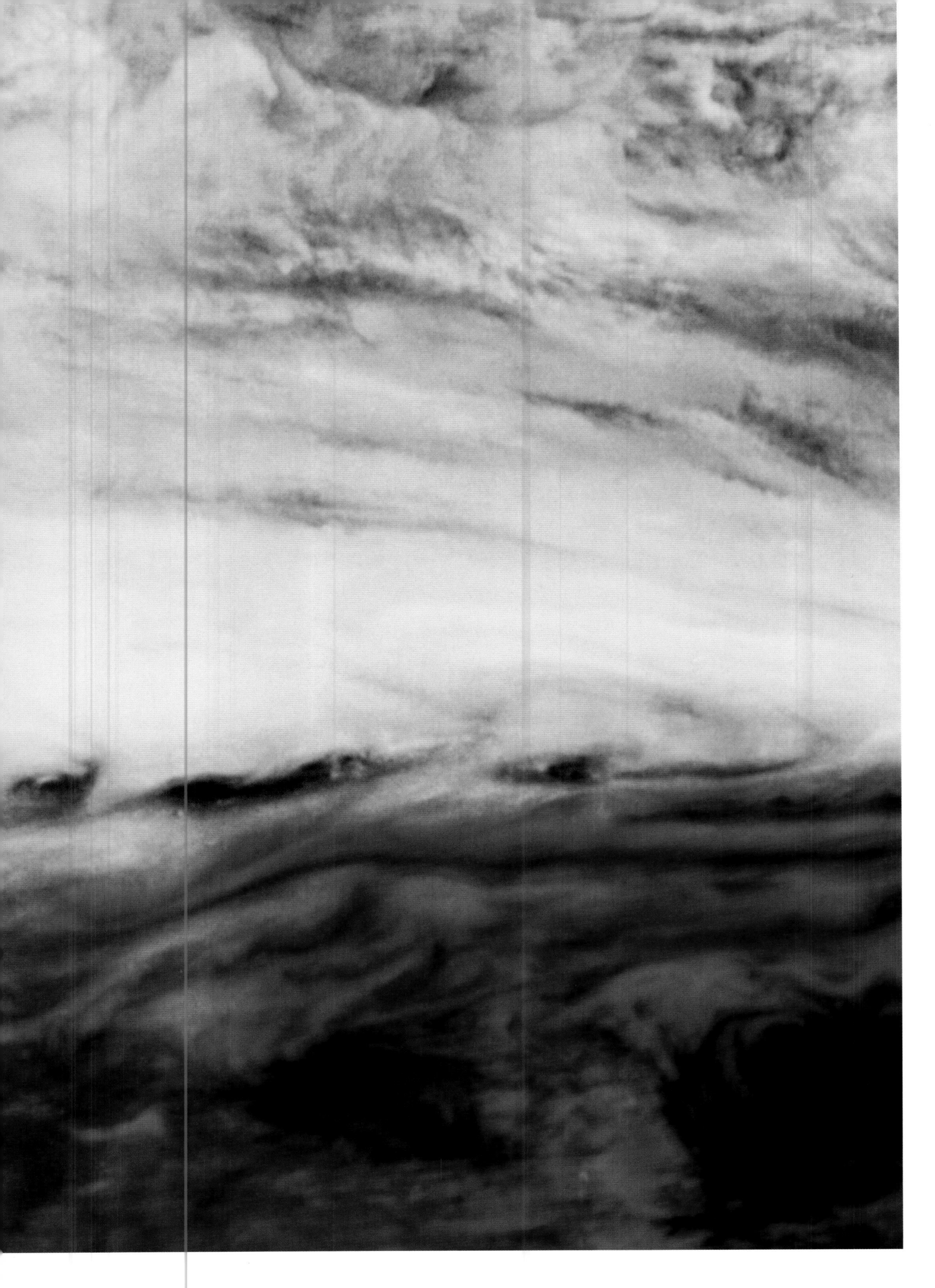

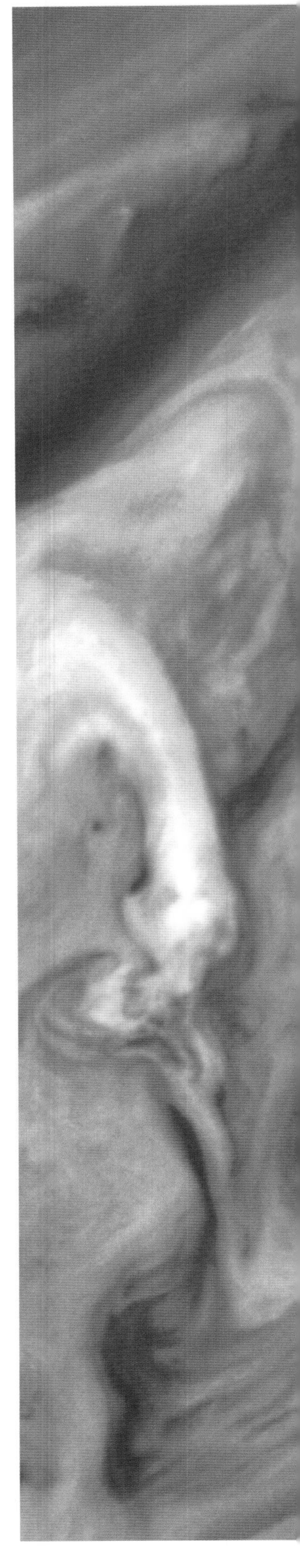

Jupiter's atmosphere

IN THIS GALILEO SPACE-PROBE IMAGE TAKEN NEAR JUPITER'S EQUATOR, a belt and a zone sit in close proximity. These regions are characterized by extremely high winds: in places, they blow at a steady 311 miles per hour/500 kph. Often, the winds in a zone blow in the opposite direction from those in a belt. This makes for enormous turbulence, causing spots and eddies to form. The violent atmosphere is also a breeding ground for enormous flashes of lightning, which break out dramatically on the nighttime side of the planet.

Jupiter: Red Spot

JUPITER'S TOURIST ATTRACTION IS UNDOUBTEDLY THE GREAT RED SPOT. This huge vortex is a giant storm some 24,856 miles/40,000 km across—three times the size of the Earth. In this area of high pressure, upwardly spiraling winds carry gases to great heights, where they react with sunlight. The red color is due to phosphorus released by these reactions. Other, smaller spots lie close to the Red Spot, but they tend to be short lived. The Red Spot, on the other hand, has lasted for at least three hundred years. This image was captured by the Voyager 2 probe in July 1979.

Jupiter's moons: Io

LOOKING LIKE A COSMIC PIZZA, Jupiter's innermost moon—Io (above)—is alive with active volcanoes (right). Despite the freezing temperature here, the Giant Planet's gravity exerts its massive forces on the interior of this moon, heating it up and causing sensational outbursts. This Galileo spacecraft image reveals the volcanic hot spots, marked by dark lava flows. The incredible colors of Io—red, orange, and yellow—are due to sulfur deposits from the eruptions. This moon is continually transforming itself and is a "must" destination for future human missions.

Jupiter's moons: Ganymede ▷

THE LARGEST MOON IN THE SOLAR SYSTEM, Jupiter's Ganymede—at 3,270 miles/5,262 km across—is bigger than the planets Mercury and Pluto. Unlike the moons closer to the Sun, Ganymede is made largely of ice instead of rock. Its crust is heavily cratered, scarred from intensive bombardment during the early days of its existence, when cosmic debris rained down on these fledgling worlds. But the surface is also crisscrossed with complex ridges and grooves—signs of more recent activity, which may have a volcanic origin.

Jupiter's moons: Europa ▽

SMOOTHER THAN A BILLIARD BALL, Europa is the smallest of Jupiter's major moons (although it's very similar in size to our own Moon). Europa's surface is brilliant white and appears to be coated by a layer of cracked ice. In this color-enhanced Galileo spaceprobe image, the brown filaments are rocky ruptures—which extend to 1,864 miles/3,000 km in length—while the blue-and-white regions are ice. Researchers believe that the thick ice crust may conceal a deep ocean of water, warmed and pummeled by Jupiter's mighty gravitational forces. Some scientists even speculate that it could harbor exotic aquatic life-forms.

Saturn: "Dragon Storm" ▷

THE SECOND-LARGEST PLANET IN THE SOLAR SYSTEM, Saturn is a massive and turbulent world. This image—captured by the Cassini probe—reveals an atmospheric complex nicknamed the "Dragon Storm" (bright orange) just above center right. The convoluted features of this long-lived cloud formation flare up from time to time to generate thunderstorms and lightning, just as we experience during storms on Earth. At the right is the edge of the planet's ring system.

Aurora on Saturn

RING-WORLD SATURN IS EVERYONE'S CHILDHOOD FANTASY OF A PLANET: a vision straight out of the covers of a science-fiction magazine. This beautiful Hubble Space Telescope image of Saturn captures its rings in all their glory. They are so wide that they could stretch nearly all the way from the Earth to the Moon—although they are only around ⅔ mile/1 km thick. The blue region at Saturn's south pole is an aurora: an eerie glow caused by electrically charged particles from the Sun colliding with the planet's magnetic field, which is strongest in the polar regions.

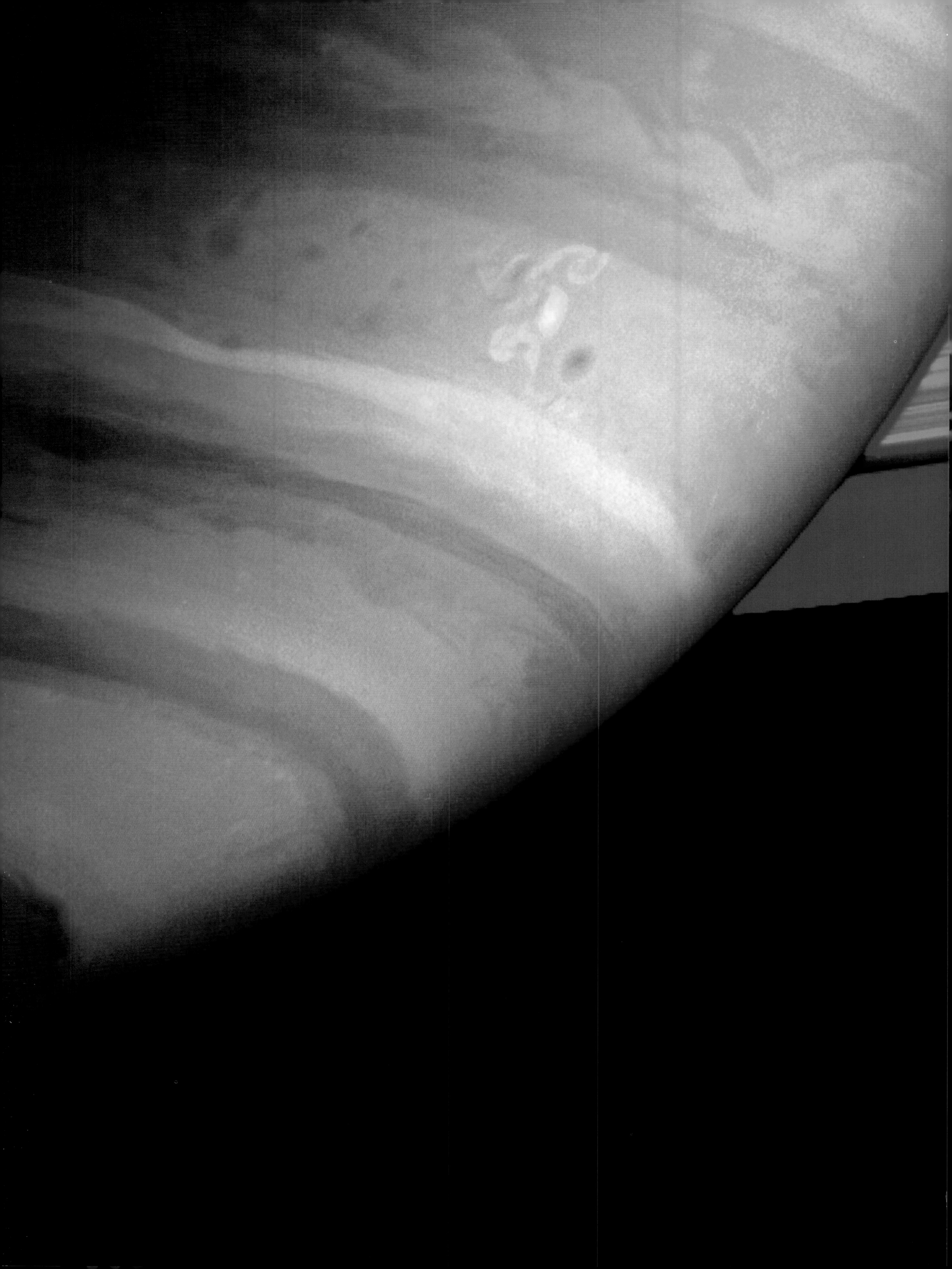

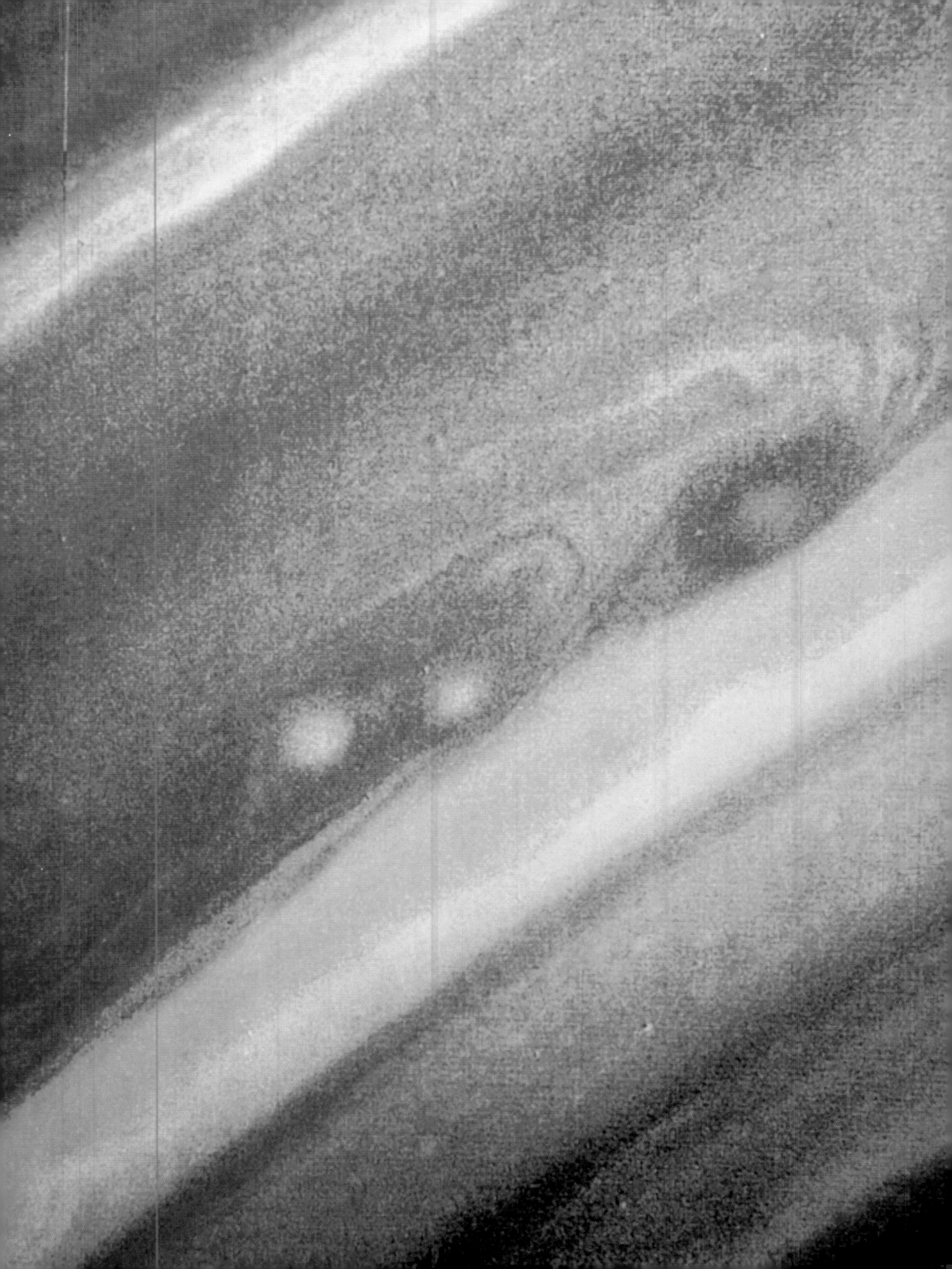

◁ Saturn: clouds

LIKE JUPITER, SATURN IS VIRTUALLY ALL GAS—but, unlike its vivid neighbor, its cloud features are blurred by a planet-wide haze. This color-enhanced image from space probe Voyager 2 in 1981 rips the haze away and reveals details of its churning atmosphere. The yellow spots are storms, while the large, reddish, elongated spot to the right of them is an anticyclone. At 3,107 miles/ 5,000 km long, it is nearly half the size of the Earth. Saturn has some of the fastest winds in the solar system, gusting to 1,118 miles per hour/1,800 kph at the equator.

Saturn's rings

WONDERS OF THE SOLAR SYSTEM: Saturn's rings, seen in 2005 by the Cassini spacecraft. The rings are not solid but comprise millions of chunks of ice ranging in size from ice cubes to refrigerators. There are also countless individual rings, as this photograph shows. This enhanced color image reveals the size of the fragments making up the rings: blue regions contain more small particles, while in red areas the fragments are larger. The rings were probably created relatively recently, as a result of a moon being torn apart by Saturn's gravity.

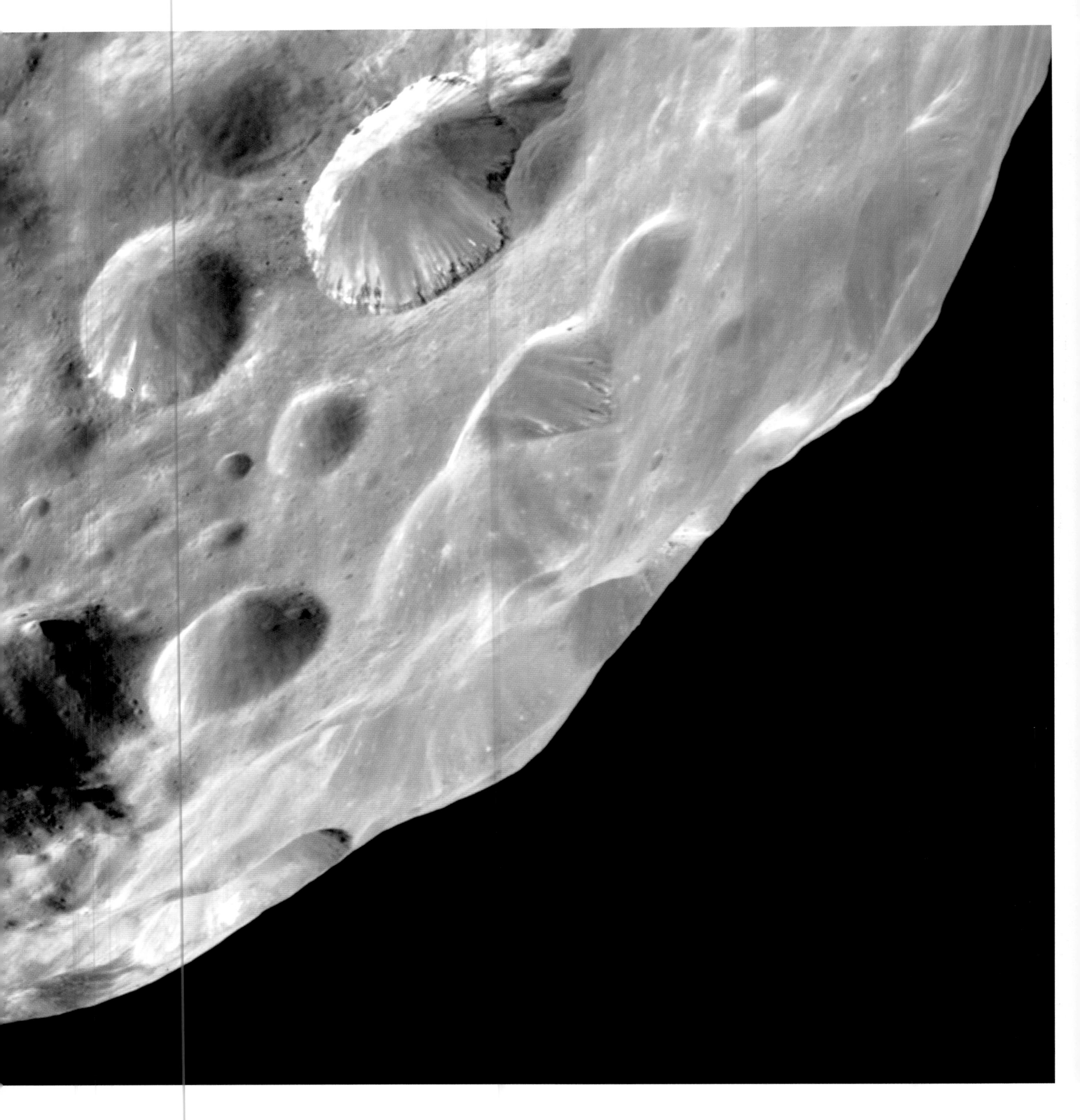

Saturn's moons: Phoebe

MEASURING 137 MILES/220 KM ACROSS, Phoebe is Saturn's outermost major moon and takes eighteen months to orbit the planet. On its way to the ring world, the Cassini probe swung by to grab this close-up view of a world pockmarked with dramatic craters gouged out by meteorite impacts. Although Phoebe is dark, this image reveals bright areas (center top and right). Researchers believe that—like most of Saturn's moons—Phoebe is actually made of ice, but it is coated with dark deposits. The current total of moons orbiting Saturn stands at forty-seven, two of which were discovered by Cassini.

Saturn's moons: Titan

SATURN'S APTLY NAMED TITAN IS THE SECOND-LARGEST MOON in the solar system and the only one to possess a dense atmosphere. Made largely of nitrogen—like Earth's—this atmosphere is orange in color and completely opaque. Not until January 2005 did we know what lay below the clouds of Titan. It was then that the European Huygens lander probe, piggybacked on Cassini, parachuted down to the surface. It discovered a rocky world with what appeared to be the relics of seas and oceans of liquid methane. These could fill up again—and some scientists speculate that, as the Sun grows larger and looms closer to Titan, Saturn's biggest moon could develop primitive life.

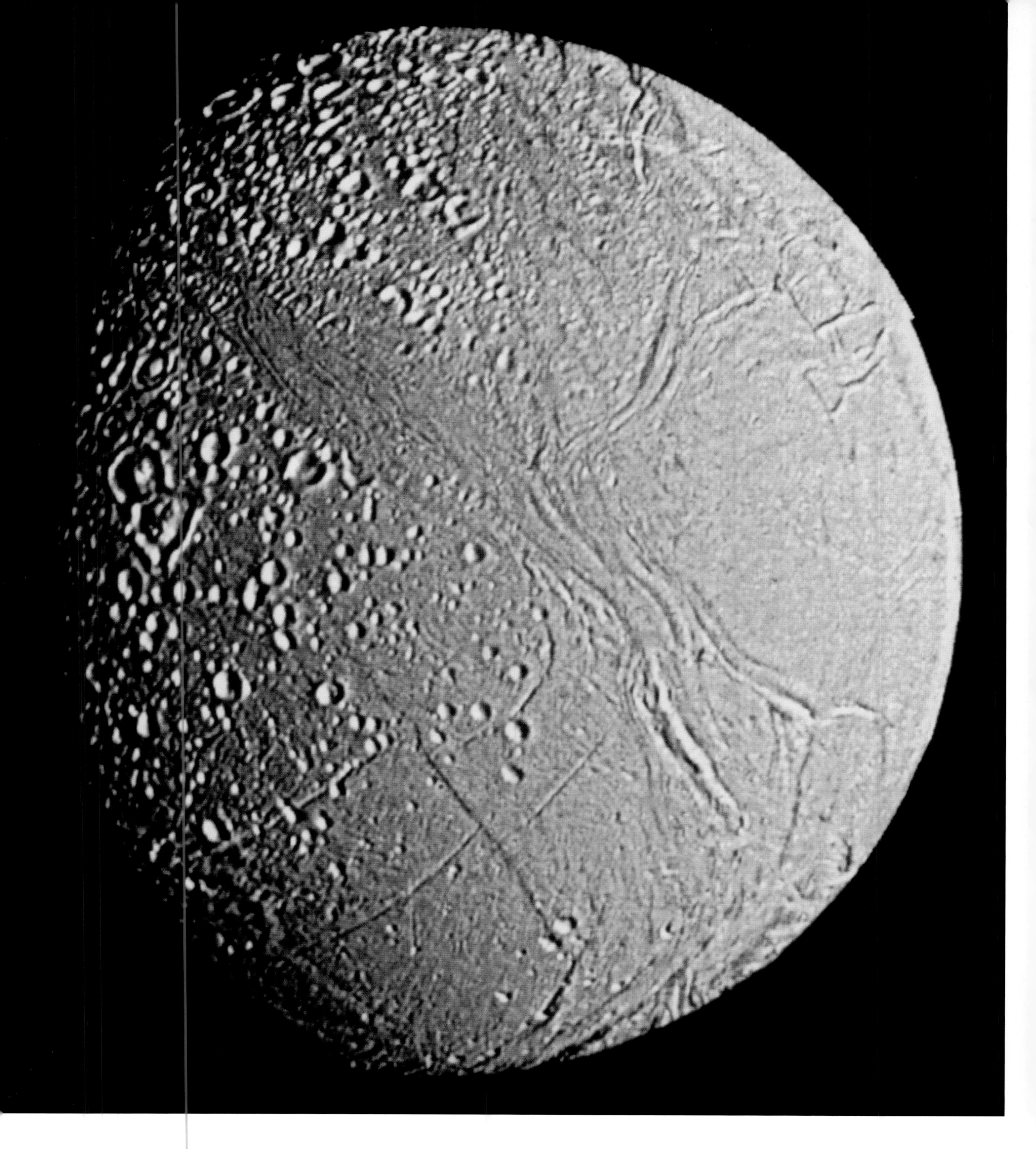

Saturn's moons: Enceladus

THE SHINIEST BODY IN THE SOLAR SYSTEM—it reflects nearly 100 percent of the sunlight falling on it—Saturn's moon Enceladus was discovered by William Herschel in 1789 (he also found the planet Uranus). Icy Enceladus, which is 311 miles/500 km across, is heavily cratered. But it also boasts grooves and canyons similar to those on Jupiter's moon Ganymede. Researchers suspect that these are recent features produced by Saturn's gravity heating up the core of Enceladus and initiating volcanic activity. Material erupted from Enceladus forms a very faint outer ring around Saturn.

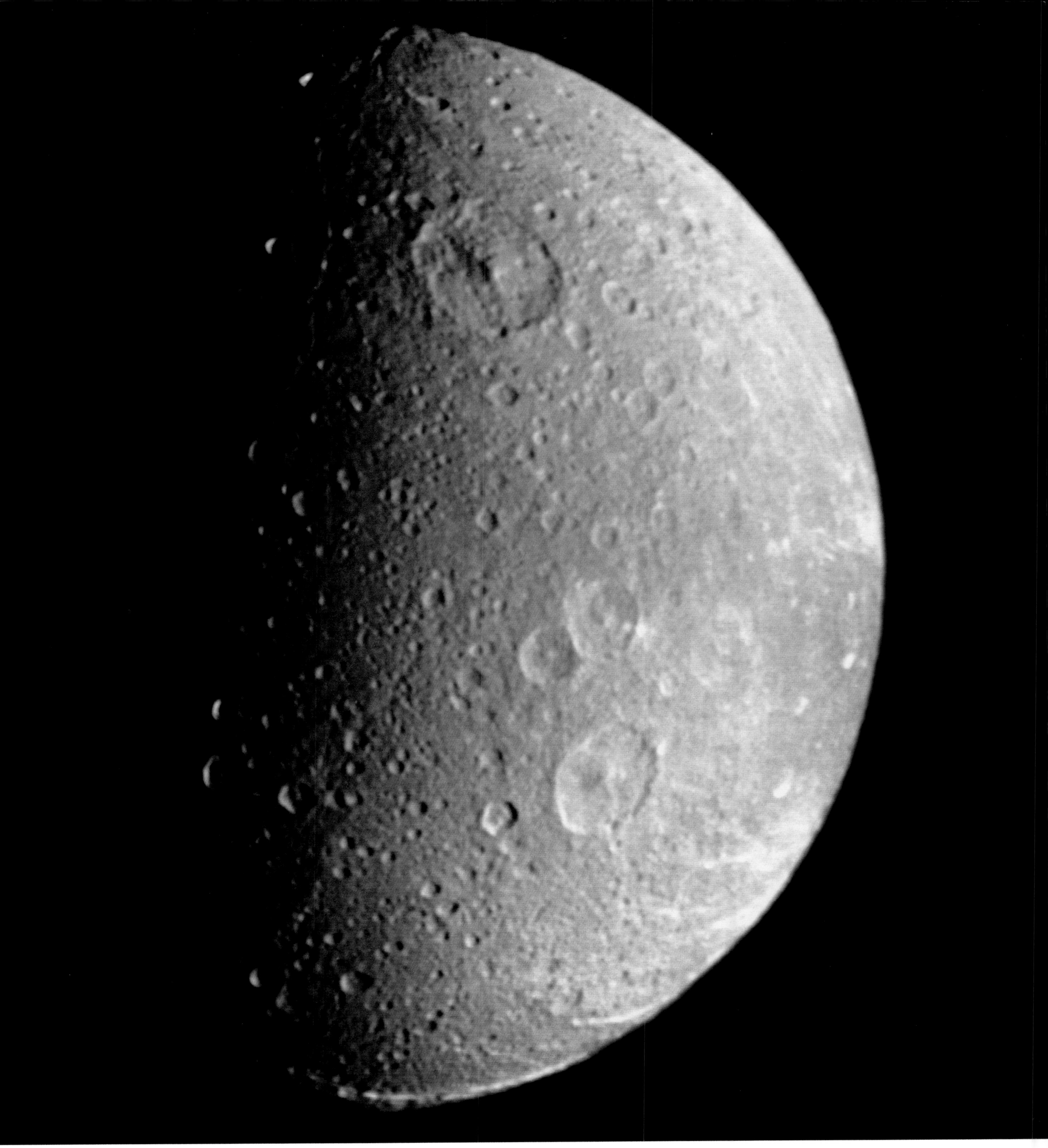

Saturn's moons: Dione

IN 1980, VOYAGER 1 CAPTURED THIS IMAGE OF DIONE, Saturn's fourth-largest moon (after Titan, Rhea, and Iapetus). Measuring 696 miles/1,120 km across, Dione bears the scars of ancient cratering. But it, too, is an active moon. Visible in this image (top left) is a long groove. The latest data from the Cassini space probe reveal features like this to be fractures and canyons formed—presumably from internal activity—in the not-very-distant past. Apart from Titan, Dione is the densest of Saturn's moons, indicating that it must be made up of rock, as well as ice.

Crescent Uranus

URANUS WAS THE FIRST PLANET TO BE DISCOVERED since the five known planets of antiquity. Barely visible to the unaided eye, Uranus was found by the musician and amateur astronomer William Herschel with his homemade telescope in 1781. Its discovery doubled the size of our solar system. Blue-green Uranus is another gas giant like Jupiter and Saturn, but it is only four times the size of the Earth. It is encircled by a set of slender rings and a family of moons. This haunting image was taken by the Voyager 2 spaceprobe in January 1986 as it left the vicinity of Uranus bound for Neptune.

Composite Uranus and three moons

FORTY-SIX MILLION MILES/74 MILLION KILOMETERS FROM URANUS, Voyager 2 captured this image of the distant world on January 24, 1986. Voyager is looking directly at the planet's bright south pole. Uranus has the distinction that it is so tilted—possibly as a result of a planetary collision—that it effectively orbits the Sun on its side. Three of the planet's twenty-seven moons are visible in the image. Miranda is on the right, closest to the planet; Ariel is next out, on the top right; and Umbriel is on the lower left. Unlike most of the moons in the solar system, which are named after mythological figures, those circling Uranus are named after characters from works by William Shakespeare and Alexander Pope.

Uranus: Hubble image ▷

TO LOOK AT, URANUS (TOP) IS THE BLANDEST PLANET IN OUR SOLAR SYSTEM—a hazy blue disk with virtually no cloud features. The chemicals acetylene and methane in the atmosphere are the culprits to blame for obscuring the details. But look at Uranus through a different wavelength filter, and a whole new world emerges. The image below, taken by the Hubble Space Telescope at infrared wavelengths in August 2003, reveals a trio of orange clouds. The Hubble has found at least twenty clouds on Uranus, which circle the planet at a speed of 311 miles per hour/500 kph.

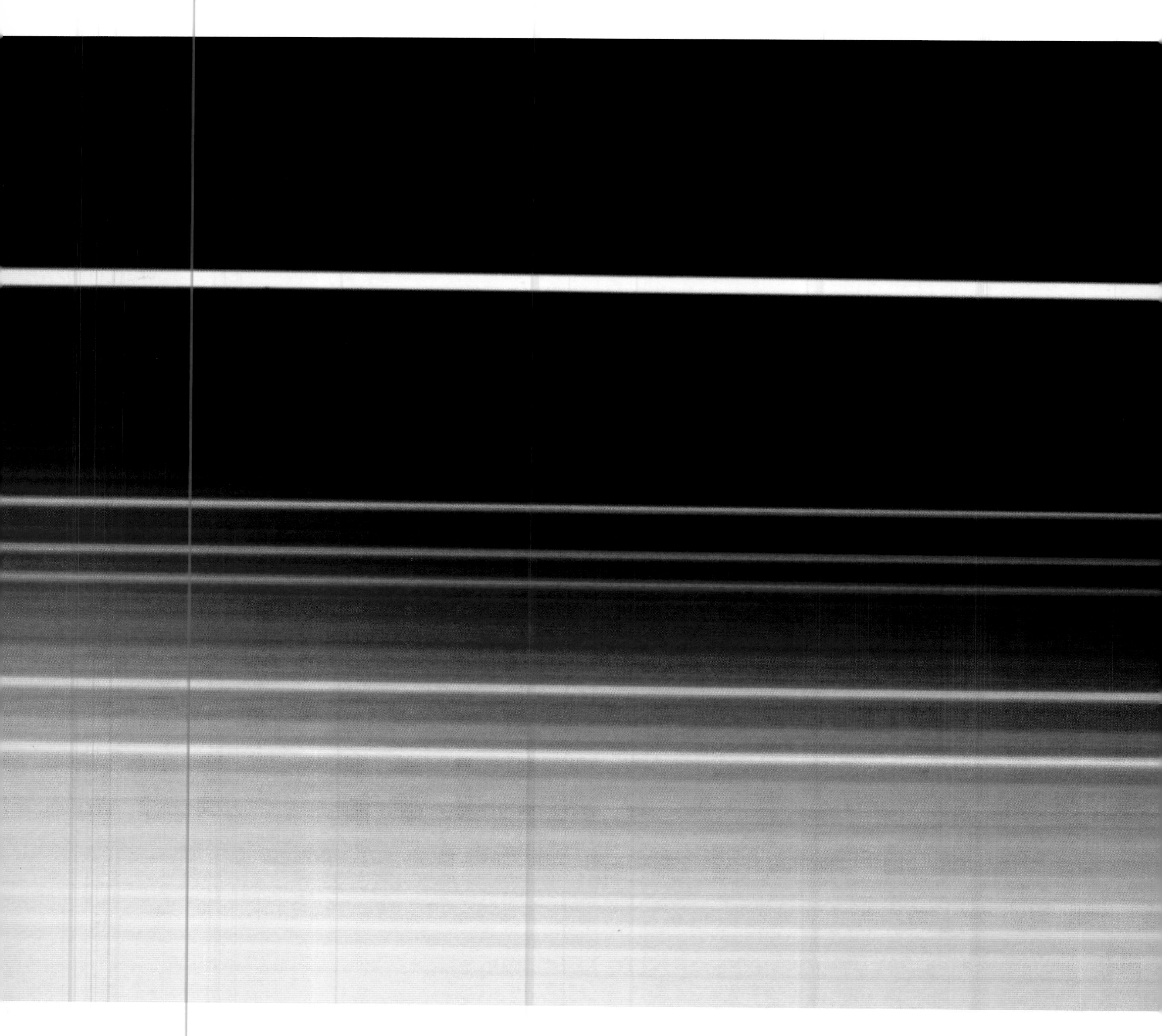

Uranus: rings

IN 1977, URANUS PASSED IN FRONT OF A STAR—and astronomers were astonished to find that the star's light blanked out nine times before and after the "occultation." The only possible conclusion was that the planet is surrounded by a set of incredibly narrow rings. Spaceprobe Voyager 2 captured the rings here on camera in January 1986. Now there are known to be at least thirteen rings. One of the outermost rings—the Epsilon Ring, 31,691 miles/51,000 km from the planet's center—is kept on the straight and narrow by the gravity of two "shepherding" moons, Cordelia and Ophelia. The rings are very dark and probably made of boulders rather than ice.

Neptune and Triton

A WONDERFULLY EVOCATIVE IMAGE OF NEPTUNE and its major moon Triton, taken by Voyager 2 as it began its long journey away from the safe sanctuary of the solar system toward the unknown depths of interstellar space. Neptune has at least thirteen moons circling it. Triton—at 1,678 miles/ 2,700 km across—is three-quarters the size of our Moon but has an unusual claim to fame: it orbits its world in the opposite direction from Neptune's other moons. Astronomers believe that it was captured by the planet after a collision with another moon.

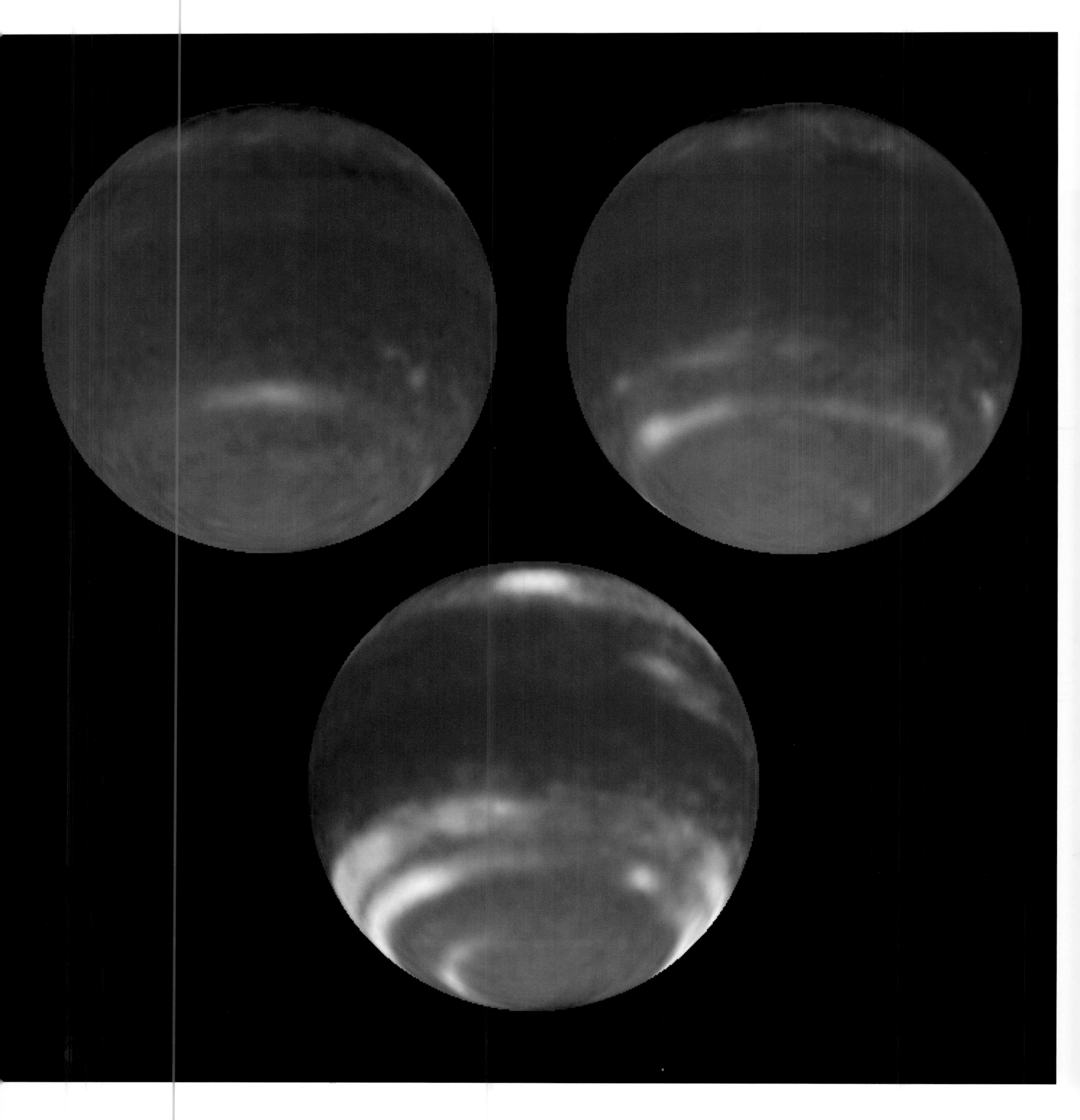

Neptune's changing seasons

THE HUBBLE SPACE TELESCOPE CAPTURED THESE IMAGES of Neptune through the seasons. Just as Earth has seasons—a result of its axis being inclined to its orbit, so sunlight doesn't fall on the planet equally at all times—the same is true of Neptune. These images were taken in Neptune's southern hemisphere in 1996 (left), 1998, and 2002 (right). The changes in the cloud patterns are thought to be due to the arrival of spring. Whatever the explanation, what is certain is that Neptune has extreme weather. The planet has the fastest winds in the solar system, gusting at speeds up to 1,243 miles per hour/2,000 kph.

Neptune's moons: Triton

NEPTUNE'S LARGEST MOON, TRITON—seen in this Voyager 2 image—is one of the coldest bodies in the solar system, yet it boasts erupting "ice" volcanoes. The dark streak is a plume of material ejected from an active vent, extending some 62 miles/100 km downwind (the moon has an extremely thin atmosphere). The ejected matter is a mix of methane, nitrogen, and dust particles. Elsewhere, the surface is furrowed with ridges and valleys, thought to be a result of alternating freezing and thawing. The moon's pinkish color is probably due to methane frost.

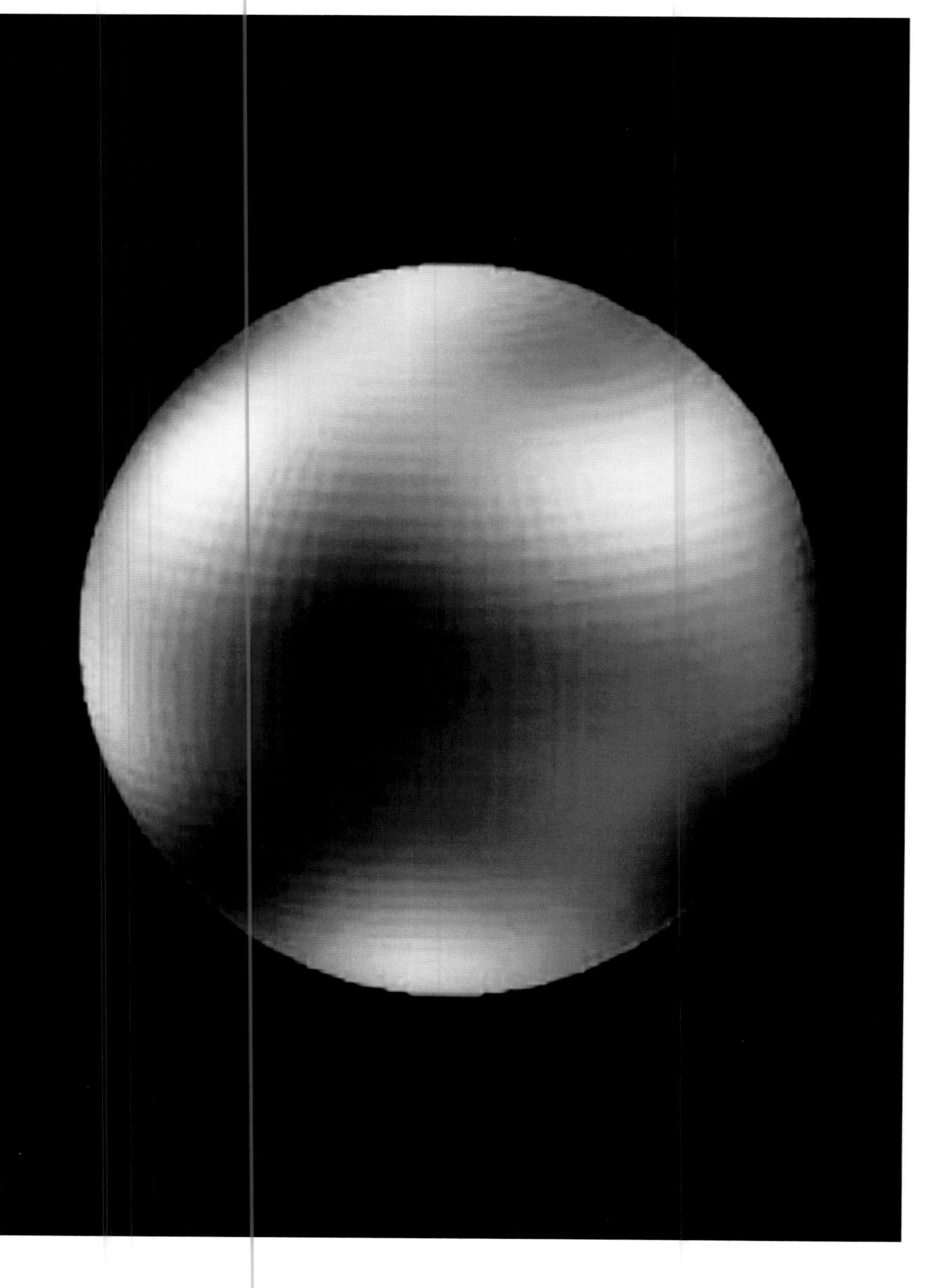

Pluto

IS IT A PLANET? This is the big controversy among astronomers at the moment. Discovered in 1930, Pluto is smaller than our Moon. And now it appears to belong to a plethora of planetesimals beyond the orbit of Neptune—and one of these (nicknamed "Xena") is actually bigger than Pluto. So is it a stand-alone planet—or just one of a swarm of "trans-Neptunian objects?" Pluto appears to be very similar to Triton, with a surface covered by methane ice. This Hubble image reveals large differences in brightness across its surface, including a large northern polar cap.

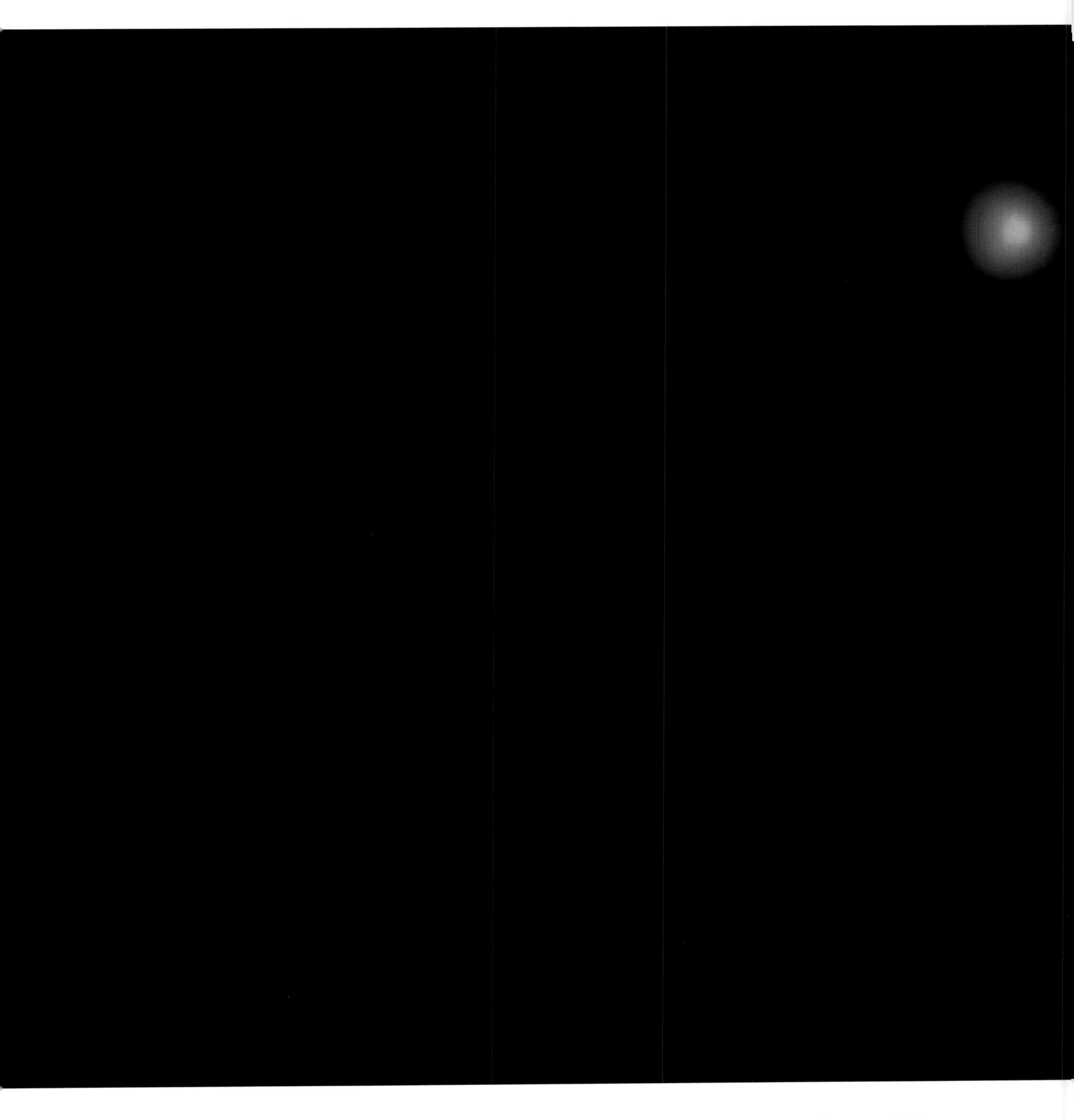

Pluto and Charon

PLUTO'S MAJOR MOON, CHARON, WAS DISCOVERED IN 1978 and is half the size of its companion world. This image, taken in 1994 by the Hubble Space Telescope, reveals the disks of the two bodies for the first time. In 2005, Hubble discovered two new, smaller moons in orbit. Currently, NASA's New Horizons space probe is bound for the Pluto system, where it will arrive in June 2015. It will also explore some of the other trans-Neptunian bodies, which are believed to be in the millions. The mission is expected to give us new insights into this twilight zone of the solar system.

When Beggars die, there are no comets seen
The heavens themselves blaze forth the death of princes

THUS SPOKE WILLIAM SHAKESPEARE in his play *Julius Caesar*, recognizing that comets have a profound role to play in our psychology, as they have acted as portents to humanity since the earliest days of antiquity. Their sudden and unannounced appearance in the skies has caused terror over the centuries—not least because everyone assumed that comets were atmospheric phenomena and might collide with the Earth.

And they were not wrong. Although comets actually live in space, they—and their partner-bodies, asteroids—have hit our planet in the past, with grave consequences. Sixty-five million years ago, a 6-mile/10-km asteroid smashed into Earth near the Yucatan Peninsula. The pall of dust and smoke that resulted from the impact blocked off the Sun and killed off the dinosaurs. Every 100 million years, an asteroid of this size will strike—one that could wipe out humanity. Smaller bodies collide more frequently. But even a comet or asteroid just 2/3 mile/1 km across could devastate a whole city.

Comets and asteroids are leftover debris from the creation of our solar system. These smaller bodies forged together billions of years ago to form the planets and moons we know today. But the mopping-up process continues. They are still causing devastating impacts on the worlds they once made: Comet Shoemaker-Levy 9's assault on Jupiter in 1994 is a salutary example.

These building blocks of the solar system live in different places. Most of the asteroids—small, rocky bodies—nestle between Mars and Jupiter. They are fragments of a world that could never be, due to their proximity to Jupiter's disruptive gravity.

Comets live in the twilight zone of the solar system. Made largely of ice, with a small mix of rock, these lonely snowballs lead a quiet life—until one of them is knocked off its perch (possibly by the gravity of a passing star). Then the comet plunges sunward, growing hotter under the Sun's rays, until it blazes into our night skies.

But comets are not all bad news. It is almost certain that the ice that they are made of brought water to the Earth. And now that we know that they usually sail past at a safe distance from Earth, we can sit back and enjoy the beautiful spectacle of a comet, with its streaming tail stretching hazily across the sky.

Comets also provide another light show: meteors. About a dozen times a year, the Earth plows through several streams of sooty dust shed by comets. These particles plunge into our atmosphere, burning up harmlessly about 62 miles/100 km up. These shooting stars create a glorious celestial fireworks display.

But asteroids are the objects to be wary of. About two thousand of them have been identified as crossing the Earth's orbit, and one of them, one day, could strike—with catastrophic consequences. Currently, astronomers are identifying them and calculating their orbits. Space scientists are working on strategies to deflect an asteroid, if ever one was to threaten us. At least we have technology on our side—unlike the dinosaurs.

4 COMETS AND ASTEROIDS

Leonids

TWO METEORS STREAK ACROSS THE SKIES of the southern hemisphere, ending their lives in a blaze of glory. These tiny grains of interplanetary debris stream into our atmosphere at speeds of up to 43 miles per second/70 km/s, where intense friction causes them to burn up. In this long-exposure image, the stars are elongated into trails as they circle the Earth's south pole. The fuzzy objects in the photograph are the Milky Way's two closest companion galaxies: the Large Magellanic Cloud (top), and the Small Magellanic Cloud (below).

Previous pages | Meteor plunging to Earth

AT A DISTANCE OF 62 MILES/100 KM ABOVE THE DESERTS of southwestern United States, a Leonid meteor plunges to its doom. This tiny fragment from Comet Tempel-Tuttle—only a fraction of an inch across—streaks past the Pleiades Star Cluster in a blaze of glory. Meteors are harmless, but their bigger cousins, comets and asteroids, could pose a threat to life on Earth.

Leonid meteors

YOU CAN SEE AT LEAST FOUR METEORS in this long-exposure photograph, all of them emanating from the same cloud of cosmic dust. Several times a year, the Earth plows through dust particles shed by comets. This debris is from Comet Tempel-Tuttle, which produces a meteor shower around November 17, called the Leonids. As a result of perspective, the shooting stars appear to radiate from the direction of the constellation of Leo. In recent years, we have encountered dense knots in this meteor stream, resulting in thousands of shooting stars an hour.

Perseid meteor trail

METEORS CAN APPEAR AT ANY TIME—every hour, around ten flash overhead—but the best time to see them is during a meteor shower. That is when the Earth intercepts dust dumped by a comet. The most-watched shower in the northern hemisphere is the Perseids, which takes place August 8–12, when many people are out in the open on vacation. These meteors—which hurtle into the atmosphere at a rate of one a minute—are debris from Comet Swift-Tuttle. This Perseid was photographed in the skies above British Columbia, Canada.

Meteor trail and the Milky Way

OVER THE MOUNTAINS OF ARIZONA, A SHOOTING STAR from the Geminid meteor shower—visible in early December—meets its demise. As the meteor burns up, it leaves a brilliant trail. This glowing streak is created by the meteor's rapid, fiery plunge, energizing our atmosphere like gas in a neon tube. The starry backdrop here is exquisite. Our Milky Way runs from top to bottom, while the constellation of Orion lies on its side at center right. At lower right, just above the mountains, is Sirius—the brightest star in the sky.

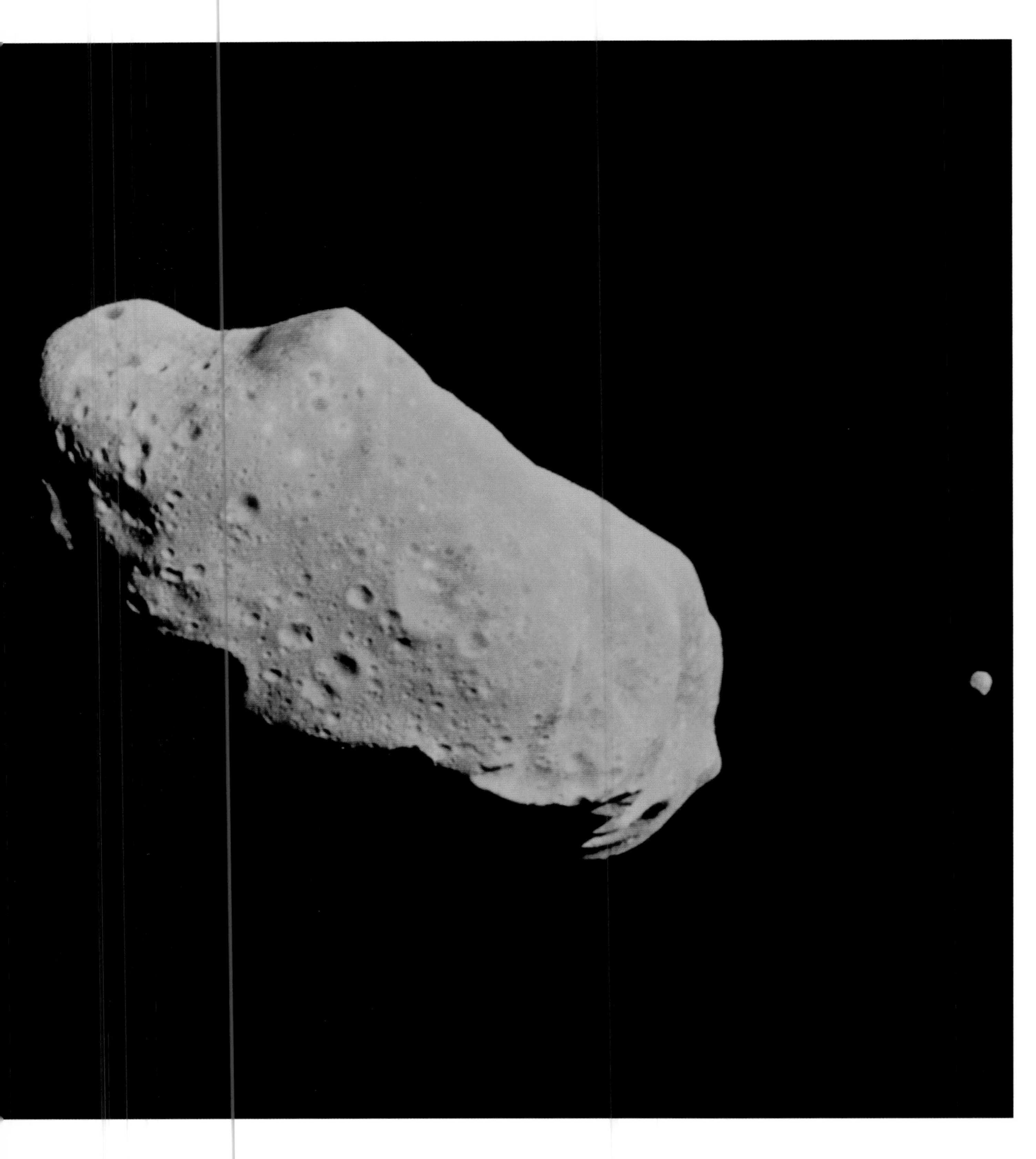

Asteroid Ida and its moon Dactyl

IN AUGUST 1993, THE JUPITER-BOUND SPACE PROBE GALILEO swung past asteroid Ida at a distance of less than 1,491 miles/2,400 km, capturing this image—which included an added bonus. The probe discovered that Ida has a tiny moon (right), now named Dactyl (after the mythological beings that lived on Mount Ida in Greece). At ⅔ mile/1 km in diameter, this tiny object is the first moon to be found orbiting an asteroid. Heavily cratered Ida is about 35 miles/56 km long and made of silicate rock. It is a member of the Koronis group: debris from a larger object (124–186 miles/ 200–300 km across) that broke up to create a family of asteroids.

Barringer Crater

A DISTANCE OF 3/4 MILE/1.2 KM ACROSS, Meteor Crater (sometimes called the Barringer Crater), is a dramatic scar punched into the sandstones of the Arizona desert. Until the early years of the twentieth century, its origin was controversial, but then Daniel Moreau Barringer—an outspoken Philadelphian mining engineer—proved that the crater must have been blasted out by a meteorite. We now know that the impact took place some 50,000 years ago. Most of the meteorite was vaporized by the blast, although drops of iron pepper the surrounding desert. The high speed of impacting meteorites mean that they cause damage out of proportion to their size; the body that struck here was just 1/20 the diameter of the crater it created.

Hubble image of an asteroid

USING THE EAGLE EYE OF THE HUBBLE SPACE TELESCOPE, astronomers have tracked down almost one hundred small asteroids moving across its field of view. In this image, the stars appear like brilliant diamonds against the awesomely dark sky. Nearly 100,000 times fainter than the dimmest stars visible to the unaided eye, the asteroid is the thin blue trail at the top of the image. Its trail is curved because of the movement of the Hubble around the Earth in this fifteen-minute exposure. The orange and blue specks are the result of cosmic rays—highly energetic particles from exploded stars—hitting the sensitive camera.

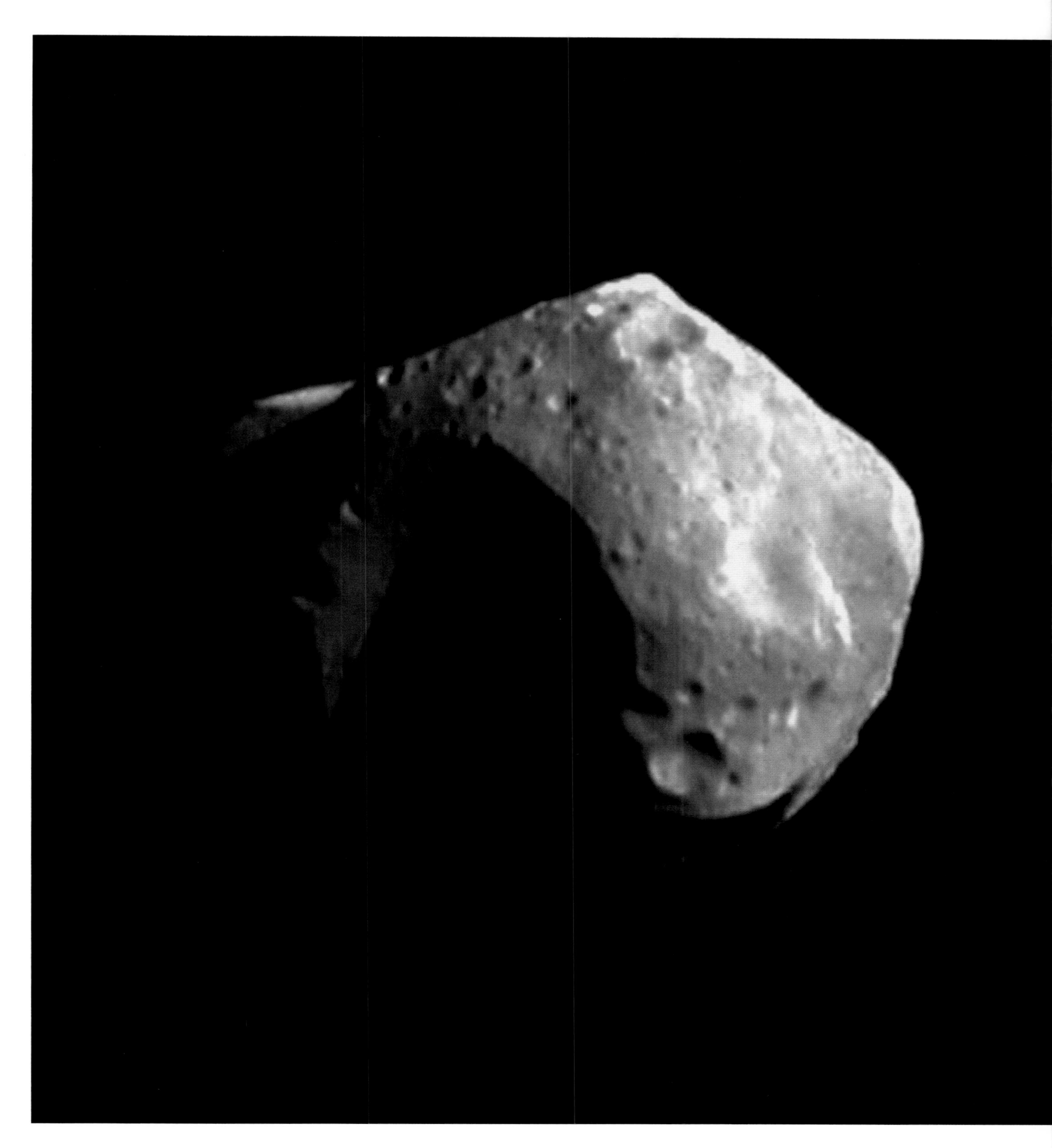

Asteroid Mathilde 253

THE NEAR (NEAR EARTH ASTEROID RENDEZVOUS MISSION) SPACECRAFT captured this image of Asteroid Mathilde in June 1997. The irregular rock measures just over 31 miles/50 km across, and it has a very different composition from Ida. Instead of being made of simple silicates, it is rich in carbon compounds. Asteroids like this—which are the majority—are extremely dark and reflect very little light from the Sun. NEAR eventually made a rendezvous with the Asteroid Eros in 2001. Renamed NEAR-Shoemaker (after renowned planetary scientist Gene Shoemaker), the probe actually landed on the asteroid's surface.

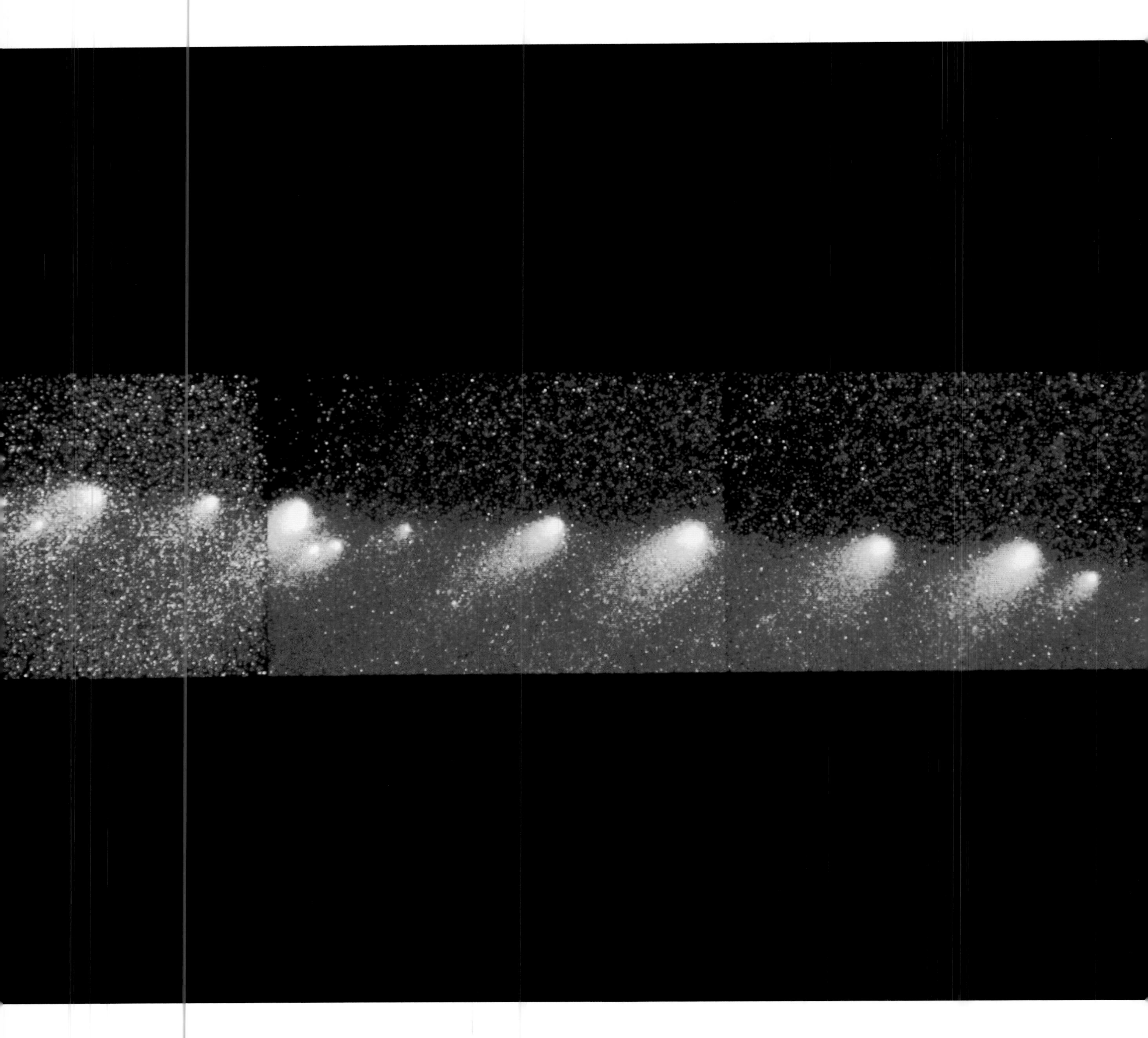

Shoemaker-Levy: the "string of pearls"

IN 1993, THE ACE COMET-HUNTING TEAM OF CAROLYN AND GENE SHOEMAKER and their colleague David Levy discovered a very strange object indeed. "It looked like a squashed comet," remembers Carolyn, the leading comet finder of the trio. Comet Shoemaker-Levy had been drawn out into a "string of pearls" by Jupiter's disruptive gravity. Broken up by Jupiter's powerful forces, the twenty-one fragments of the comet were now on a collision course with the giant planet—an event that would take place a year later.

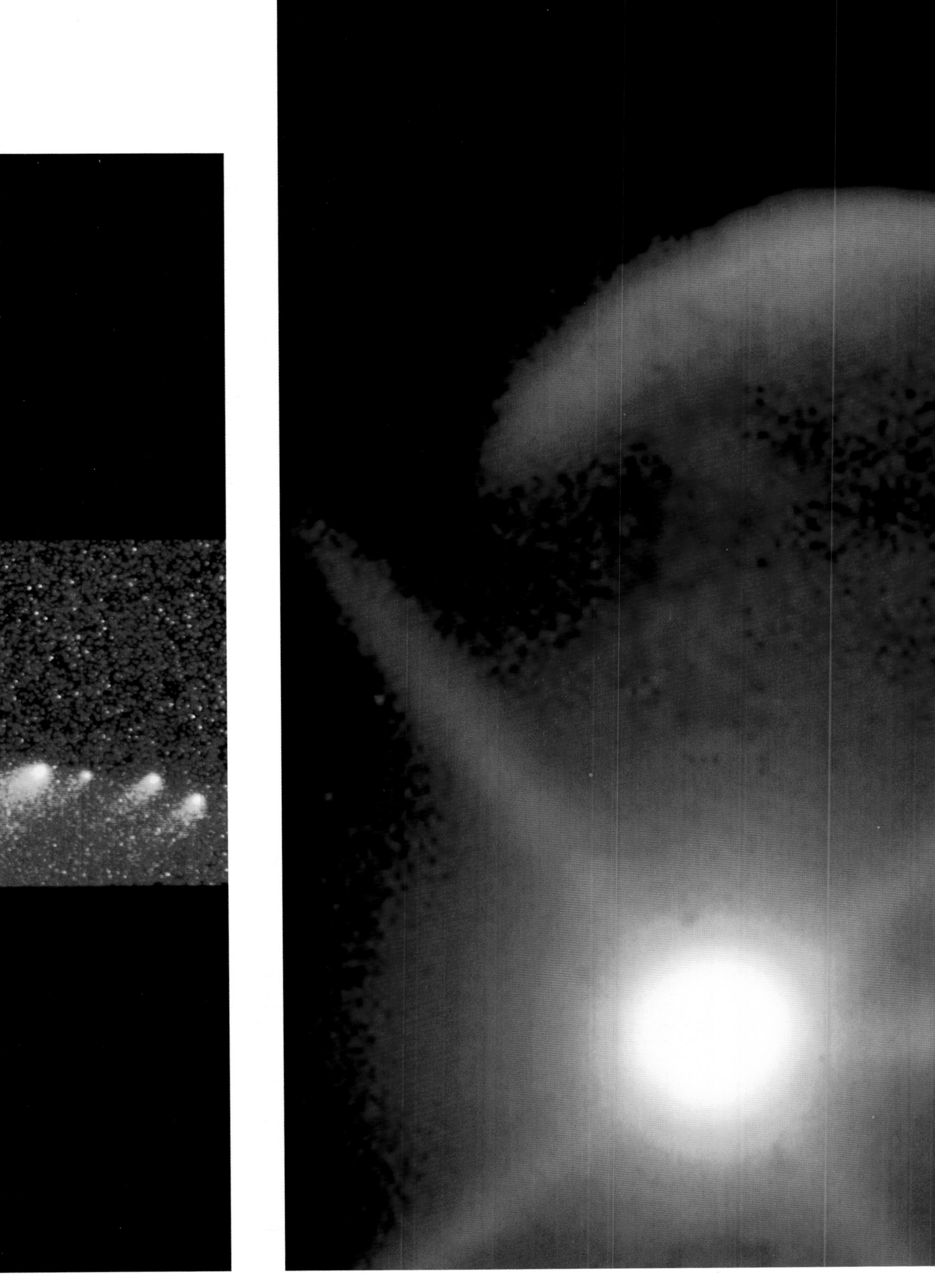

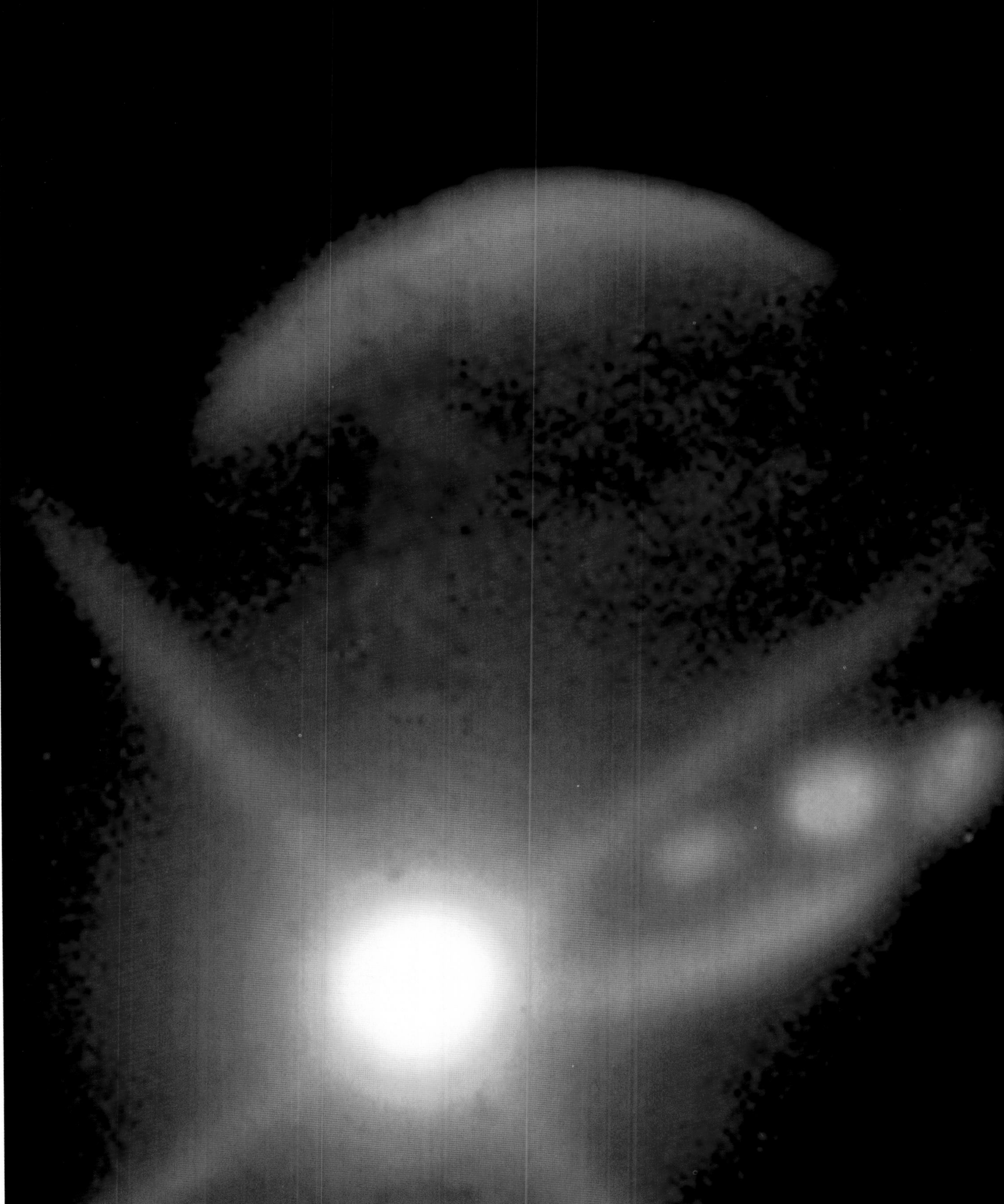

Shoemaker-Levy impact on Jupiter

JULY 1994: COMET SHOEMAKER-LEVY SLAMS INTO JUPITER at a speed of 124,000 miles per hour/200,000 kph—twenty-one times over. These colossal impacts inflicted massive damage on the planet. This image, from Mount Stromlo Observatory, Australia, shows a fireball 12,427 miles/ 20,000 km across arising from one of the first hits. In the aftermath, Jupiter was pockmarked with dark scars—which took months to disappear. The collision was a timely warning that the Earth is also vulnerable to impacts by comets and asteroids. A body 6 miles/10 km across can impact our planet about every hundred million years—leading to mass extinctions.

Comet Williams

FIRST PINPOINTED IN AUGUST 1998 by the Australian astronomer Peter Williams, this comet was photographed a few weeks later over New South Wales. Comets are commemorated after their discoverers: up to three of them can be incorporated in the comet's name if they find the object independently. This image shows Comet Williams heating up as it approaches the Sun. Its icy crust is boiling, and pressure from the Sun's outflowing atmosphere—the solar wind—is forcing the comet's gases to stream out behind in a narrow tail.

Comet West

ONE OF THE GREATEST COMETS OF THE TWENTIETH CENTURY, Comet West was an unsung hero. It was discovered in 1975 by Richard West of the European Southern Observatory in Chile. But the world was still disappointed over the faintness of Comet Kohoutek in 1973—predicted to be the brightest comet of the century—and, as a result, largely turned its back on this cosmic interloper. West was a sensationally brilliant object in the morning skies of early 1976, and it was even visible in daylight through binoculars. Astronomers have calculated that West takes 254,000 years to orbit the Sun.

Comet Ikeya-Zhang

THIS COMET WAS DISCOVERED INDEPENDENTLY BY TWO AMATEUR ASTRONOMERS in Japan and China in 2002. Ikeya-Zhang later became visible in the night sky and boasted a sensational pair of tails boiling off the comet's nucleus. The straight, blue, gas tail points directly away from the Sun, because our local star's powerful magnetic field streamlines the comet's energetic atomic particles. The reddish dust tail—originating from the solid fragments in the comet—curves lazily away into space. At the bottom of the image is one of the most distant objects visible to the unaided eye: the Andromeda Galaxy, nearly three million light years away.

Comet C/2002 T7

THIS COMET DOESN'T HAVE A REAL NAME, because it was discovered by a robot! The LINEAR (Lincoln Near Earth Asteroid Research) project—a joint enterprise between Massachusetts Institute of Technology's (MIT) Lincoln Laboratory and the United States Air Force—aims to find objects that could potentially impact the Earth. The project's technology, derived from the surveillance of hostile missiles and satellites, has now been turned into more peaceful purposes. Using automatic telescopes based at the White Sands Missile Base in New Mexico, the LINEAR program discovered this comet in 2002. It passed at a safe distance of 25 million miles/40 million km from Earth in May 2004.

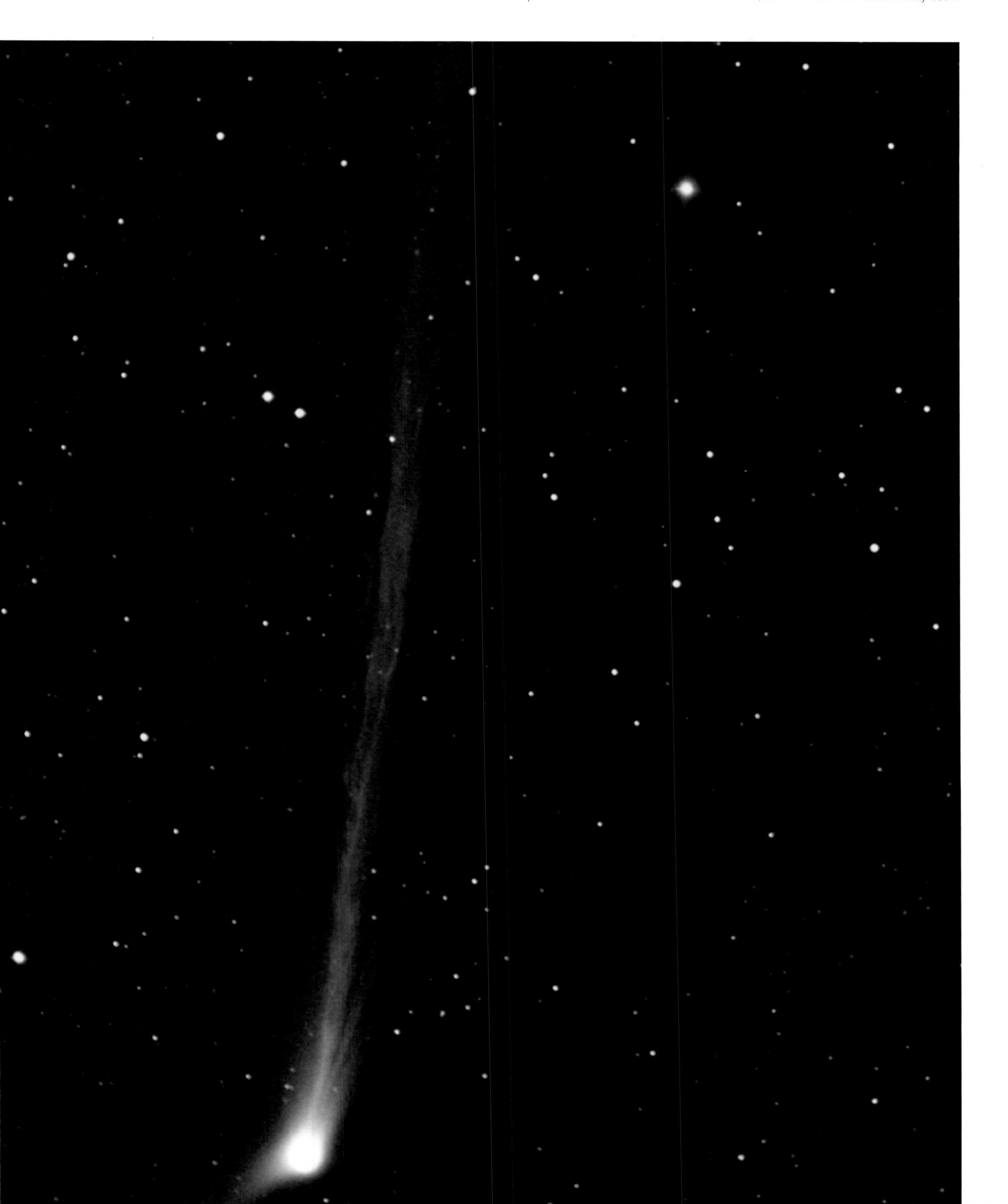

Comet C/2001 Q4

ALL AROUND THE WORLD, astronomers are on the alert for objects that could collide with our planet. The automatic telescopic survey NEAT (Near Earth Asteroid Tracking program)—based on Palomar Mountain, California—detected this comet in 2001. This particular image was captured by the WIYN (Wisconsin, Indiana, Yale, and NOAO [National Optical Astronomy Observatories]) telescope at Kitt Peak Observatory in Arizona. Between the comet and the red star on the bottom right is a small cluster of stars called Melotte 72. The huge head of gas—or "coma"—of a comet can measure up to 620,000 miles/1 million km in diameter.

Comet Bradfield

ON APRIL 12, 2004, AUSTRALIAN ASTRONOMER BILL BRADFIELD discovered his eighteenth comet. He has been searching for these fragile ships that pass in the cosmic night since 1972. The comet's long tail is beautifully visible in this image of the twilight sky of April 24. At its greatest, it extended to a length of 20 degrees—the extent of a hand—with fingers extended, held at arms' length. Bradfield, a retired rocket scientist, believes his investigations go beyond the automatic searches of NEAT and LINEAR. "These are based in the northern hemisphere, and can't see what I call 'deep south.'"

Hale-Bopp with Aurora Borealis ▷

COMET HALE-BOPP BECAME A SENSATIONAL SIGHT in the skies of April 1997 when it outshone every star except brilliant Sirius. In this image, captured in Alaska, the comet's two tails are gloriously visible against the starry backdrop. The glow below the comet is not twilight: it is the Aurora Borealis. In regions near the poles, like Alaska, the Earth's magnetism channels the electrically charged particles from the Sun downward. They then collide with the atmosphere to create a light show rivaling a Fourth of July display. The green and red colors are haunting; so too are the incredible shifting curtains and rays of this beautiful phenomenon.

Space Shuttle image of Comet Hale-Bopp

VIEWED FROM SPACE SHUTTLE COLUMBIA, the brightest of the recent comets—Comet Hale-Bopp—hovers over the Earth against the deep blackness of space. The comet was discovered independently on July 22–23, 1995, by two astronomers. Alan Hale was a professional astronomer, focusing on stars, but he was driven by a passion for finding comets (this was his first). Tom Bopp, on the other hand, was an amateur astronomer who had never hunted for comets before. At a star-party in Arizona, he just happened to point a telescope in the direction of the star clusters in Sagittarius . . . and the rest is history.

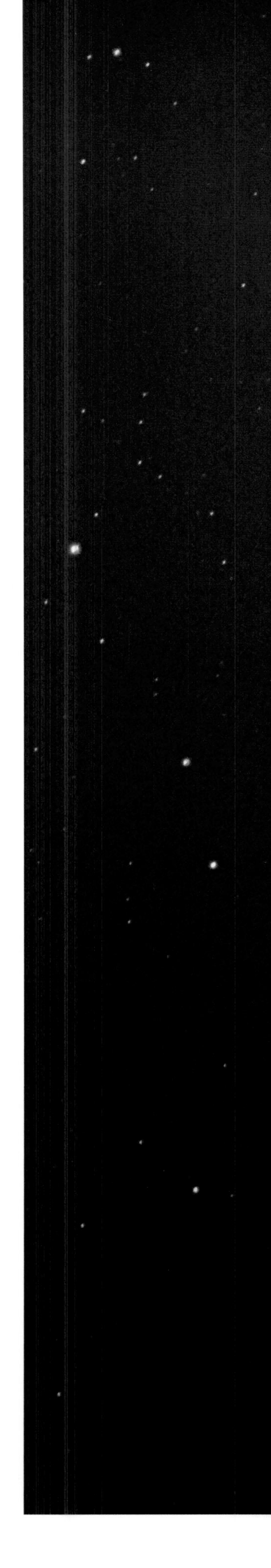

Comet Hyakutake and Arcturus

ON JANUARY 30, 1996, the Japanese amateur astronomer Yuji Hyakutake discovered a comet (his second) using a pair of powerful binoculars. It became one of the brightest comets of the late twentieth century, passing the Earth at a distance of 10 million miles/16 million km, in March of the same year. In this image, the comet boasts an exceptionally long tail, stretching across a quarter of the sky. The comet was photographed on March 26 from Australia, and the bright orange star in the sky is Arcturus, usually associated with the northern hemisphere.

Comet Hyakutake

HEADING TOWARD THE SUN—having left the vicinity of the Earth—this image of Comet Hyakutake was captured in April 1996. Its last visit to the inner solar system was around 17,500 years ago, and it won't return for at least 29,500 years. At least the comet will immortalize Yuji Hyakutake, who died suddenly in 2002 at the early age of 51. He once said: "I don't care about the naming of this comet. If many people could enjoy that comet, that is the happiest thing for me."

Halley's Comet

THIS STUNNING ARTWORK OF HALLEY'S COMET shows the cosmic interloper at its most active, with gas streaming behind in a tail millions of miles long. Its first recorded appearance was in 240 BC, and it has been plunging into the inner solar system every seventy-six years since. Every time it visits, it loses material. In fact, some of the debris streaks into our atmosphere in the form of the Orionid and Eta Aquarid meteor showers, so Halley won't last forever. Astronomers estimate that the comet will live for another 250,000 years.

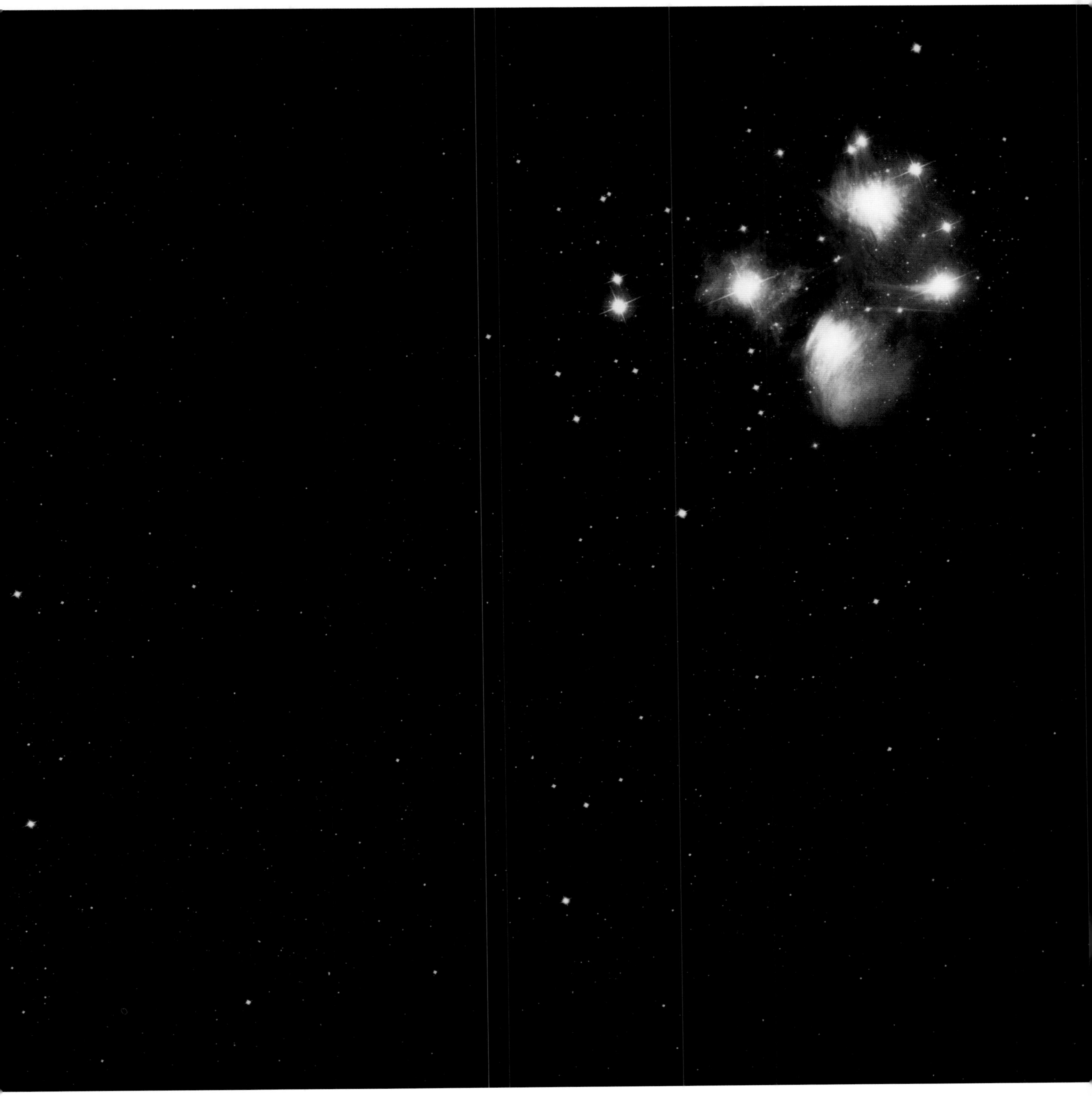

Halley's Comet and the Pleiades

CAPTURED IN NOVEMBER 1985—as Halley was on its way toward the Sun—this image from the Cerro-Tololo Inter-American Observatory in Chile shows the comet next to the dazzling young Star Cluster of the Pleiades ("Seven Sisters"). At this time, the comet was putting on its best appearance for viewers in the northern hemisphere and was relatively faint; in fact, many people confused the Pleiades with the comet. It brightened as it approached the Sun and moved to the southern hemisphere—where it was a beautiful sight. Halley returns in 2061—but will, unfortunately, be even further from Earth.

Halley's Comet in close-up

IN 1986, EUROPE'S GIOTTO SPACE PROBE SWOOPED through the coma of Halley's Comet and imaged its nucleus: the solid heart of the comet. It was a first—no one had seen a "naked" comet before. But Giotto, speeding at 155,000 miles per hour/250,000 kph, was badly damaged during the encounter by dust particles erupting from the active comet. The nucleus is a cratered, potato-shaped lump of ice and rock measuring 10 x 5 miles/16 x 8 km, coated with a material blacker than coal. Under the Sun's heat, jets of gas break through the thin crust, and the dust coating the surface streams away into space.

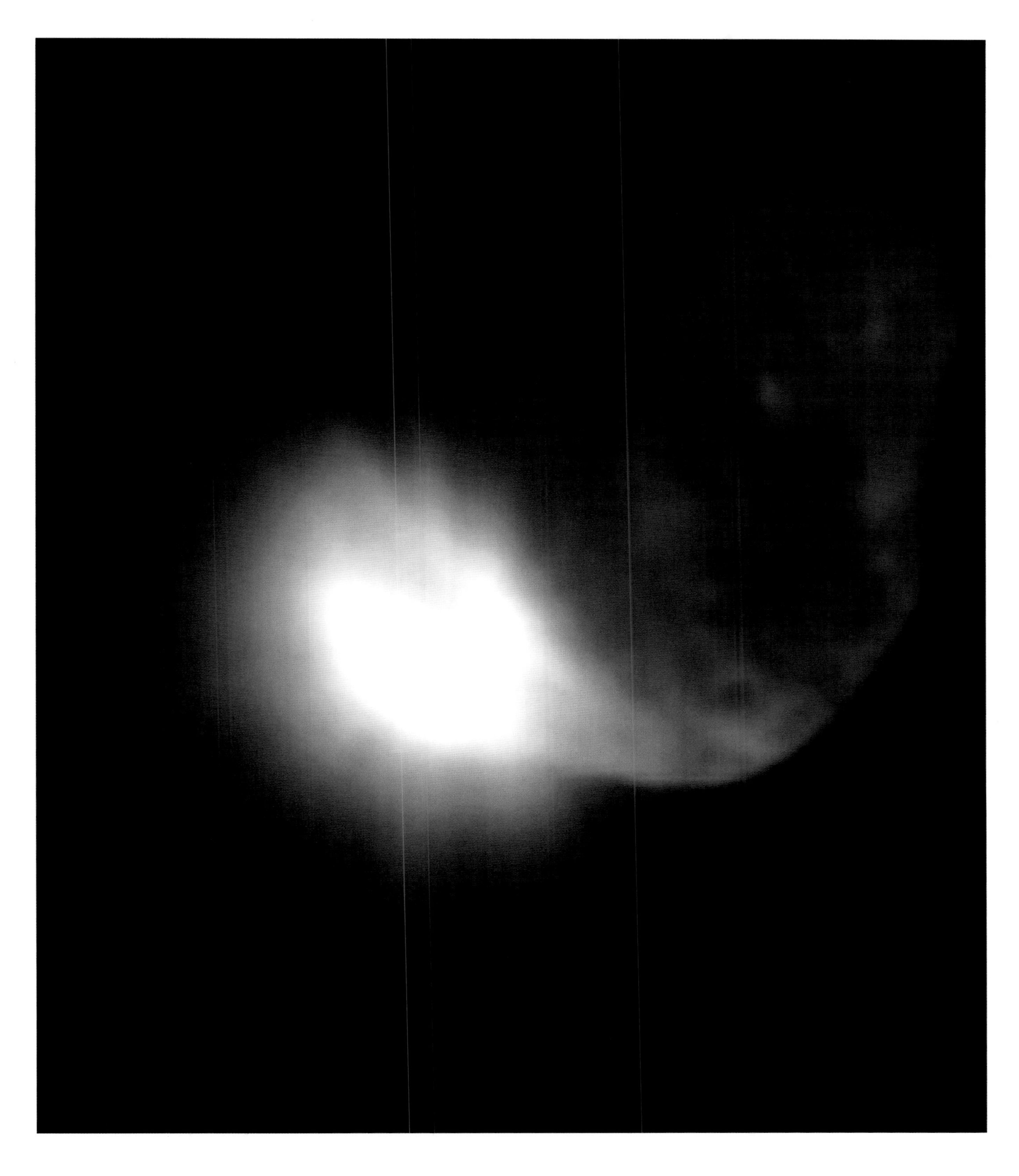

Deep Impact and Comet Tempel 1

HOLLYWOOD MEETS A COMET: on July 4, 2005, NASA's Deep Impact probe smashed into Comet Tempel 1 at a speed of over 19,000 miles per hour/30,000 kph. The probe—about the size of a coffee table—caused an immense explosion, monitored by the mother craft from a safe distance. The object of the exercise was not to cause wanton destruction but to analyze the material inside a comet's nucleus. This buried ice and rock has remained unaltered since the earliest days of the solar system and can offer astronomers clues as to what conditions were like when the Sun and planets were formed.

LIKE OUR SUN, ALL STARS HAVE A FINITE LIFETIME: they are born, they live, and they die. But their lifetimes are so long—measured in billions of years—that it is rare to see a star changing as it ages. And one of the most elusive clues to understanding how stars live is finding out how they were born.

Stars are natural nuclear fusion reactors. Made of gas, the pressures in their interiors are so intense—and the temperatures are so high—that the hydrogen gas in a star's core welds itself into helium. This reaction creates the energy that we see as starlight.

To understand star birth, we have had to resort to a lot of new technology. Now, telescopes and satellites above the Earth are able to look at the heavens in wavelengths of light we are unable to access from beneath our planet's churning atmosphere. In particular, the Spitzer Space Telescope—which homes in on heat radiation (infrared)—can penetrate the hidden recesses where fledgling stars are being created.

It all starts in the dark. Even the most brilliant stars begin their lives hidden from view, deep within vast, dark swathes of gas and dust called molecular clouds. Some of these clouds are visible to the unaided eye, silhouetted against the glowing band of the Milky Way. Now, gravity is the driver. It forces the cloud to collapse, making it fragment into thousands of smaller clumps known as protostars. These shrink further, growing hotter and hotter as the gas is compressed—like the air in a bicycle pump.

Once the central temperature of a protostar has reached several million degrees, nuclear reactions kick in. Hydrogen at the star's heart fuses into helium, which releases floods of energy. The star begins to shine: a new sun is born.

The newly born stars light up their surroundings with considerable vigor. We see these glorious maternity wards in the heavens as nebulae—clouds of gas and dust illuminated by their progeny. And they are absolutely vast. One of the nearest—the Orion Nebula, visible to the unaided eye—is fifteen light-years across. Which means that a ray of light, traveling at 186,000 miles per second/300,000 km/s, would take fifteen years to cross it. The Tarantula Nebula, in our neighboring galaxy, the Large Magellanic Cloud, is one thousand light-years in diameter.

These nebulae are the some of the most beautiful objects in the cosmos. Laced with lanes of sinuous dust and gas, and glowing with the new light from young stars, they light up our skies like celestial butterflies.

Yet as soon as stars have been born, the nebula is doomed. Young stars are violently destructive. Charged with the precocious energy of youth, they wipe out their natal material with powerful jets of gas, coupled with vicious stellar winds. Their violent birth disperses the gas into space.

But it is not the end. Fortunately, our universe has the ability to recycle its material, so that the atoms from a dying nebula will find their way into a new star.

5 STAR BIRTH

Star birth in the Gum Nebula

SILHOUETTED AGAINST THE HUGE GUM NEBULA, dark, dense clouds of sooty dust begin the process of star birth. Initially cold, these clumps begin to shrink under their own gravity, and the resulting pressure on the gas and dust heats up their interiors. Eventually they break up into dozens of protostars. The surrounding Gum Nebula is a reservoir of hydrogen gas: a breeding-ground for fledgling stars. Eight hundred light-years across and covering sixty full moon-widths in the sky, it is one of the biggest nebulae in the galaxy. It was discovered in 1953 by the young Australian astronomer Colin Gum, who tragically died in a skiing accident seven years later.

Previous pages | Omega (or Swan) Nebula

A HAZY CLOUD OF GAS AND DUST IN SPACE prepares itself for one of the greatest acts of creation: star birth. Here, in the Omega Nebula, there is enough material to make a million stars. The process has only just started, and the fledgling protostars are still cocooned in their nest. But over the next few million years, we can expect to see a blaze of activity from this stellar nursery, 5,000 light-years away.

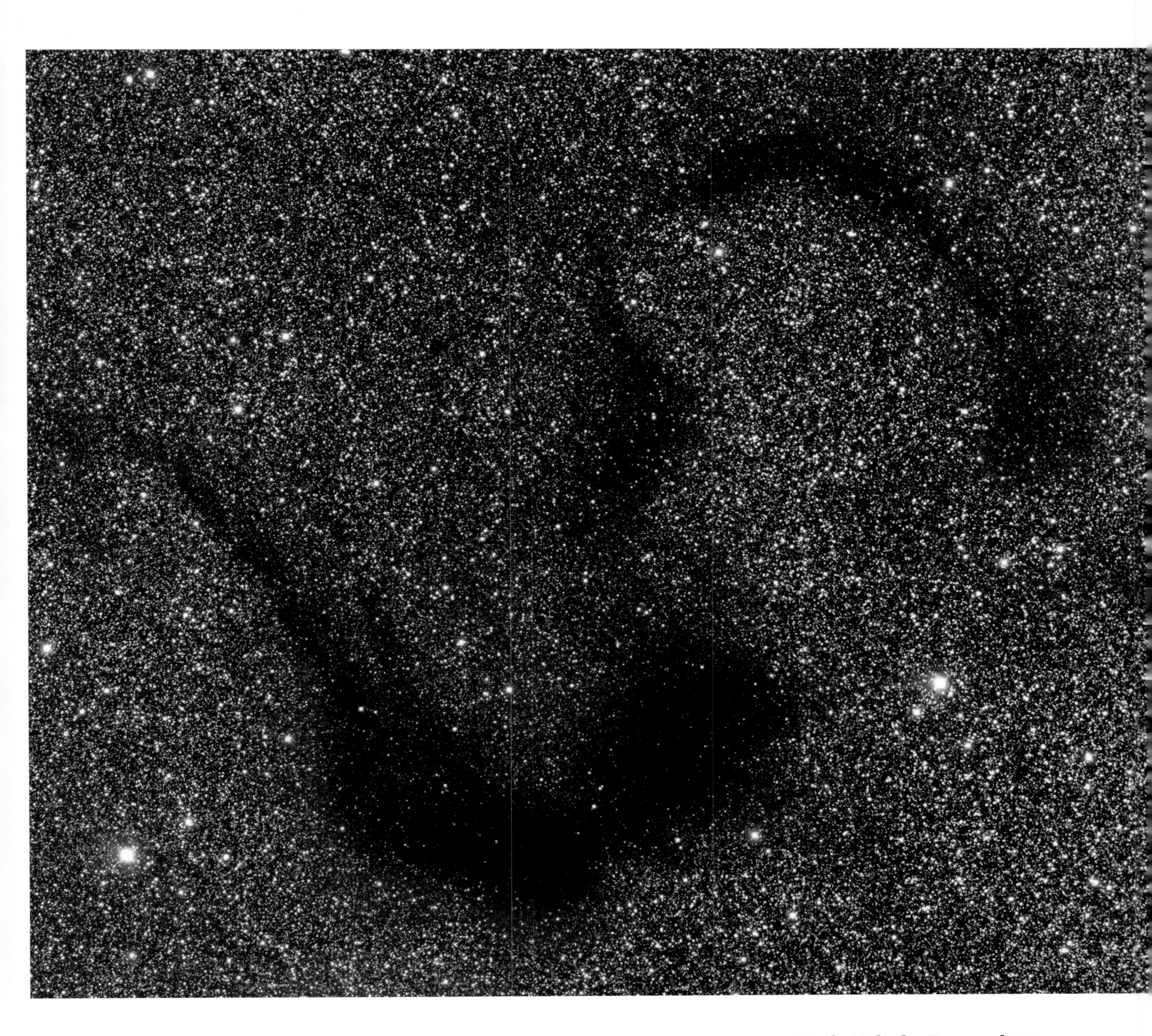

Dark Nebula Barnard 72

APTLY CALLED THE "SNAKE NEBULA," this dark cloud of gas wriggles its way across the brilliant star field behind it. Located in the obscure constellation of Ophiuchus, the officially named Barnard 72 was discovered by the American astronomer Edwin Emerson Barnard. During the late nineteenth century, he discovered 182 dark nebulae during an exhaustive survey of the Milky Way, at California's Mount Wilson Observatory. The nebula is rich in dust that has boiled off the surfaces of old, cool stars—and will later provide the raw material to create planets around new stars.

A Hubble image of star birth

A GALAXY IN TURMOIL, CAPTURED IN THIS IMAGE from the Hubble Space Telescope. Forty million light-years away, NGC 1808—a star city like our Milky Way—is undergoing a frenzy of star formation. In its central regions, pictured here, hot new stars shine blue; more mature stars are yellow. NGC 1808 is known as a Starburst Galaxy, where the rate of star birth and the activity it generates are way above the norm. The cause of this celestial chaos could be the proximity of a neighboring galaxy whose gravity is raising tides on its companion.

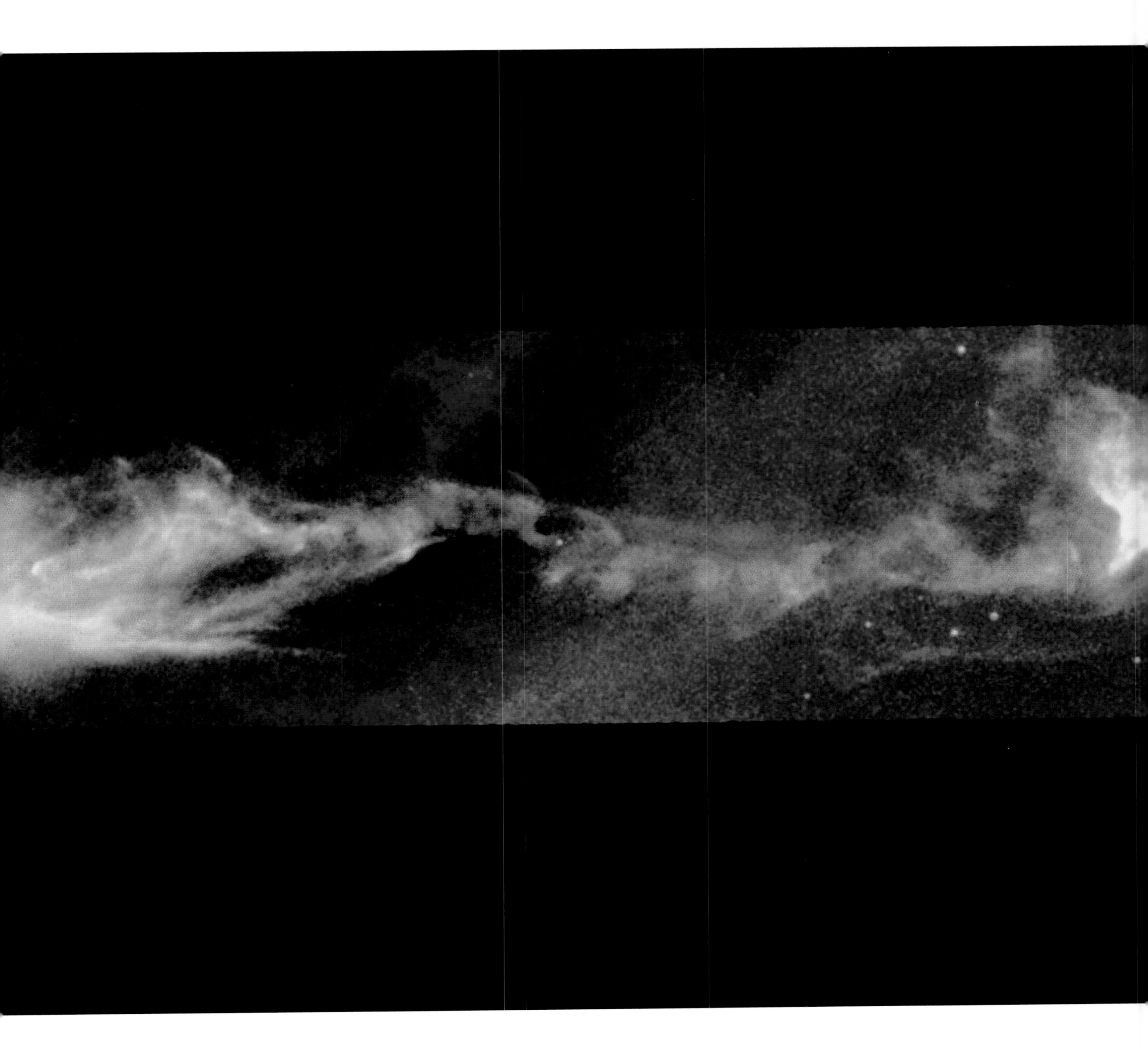

Gas from a young star

A YOUNG STAR—HH-47—EMITS POWERFUL JETS OF GAS from its poles, spanning across half a light-year of space. One of many such fledgling stars, it was named after the astronomers George Herbig (from the United States) and Guillermo Haro, who pioneered research into star birth. Haro was Mexican and devoted much of his life to building up new astronomical research institutes and educational facilities in his native country. The jets from HH-47 travel at speeds of around 186 miles per second/300 km/s. They appear wiggly, possibly because HH-47 is wobbling under the influence of an unseen companion star.

Emission Nebula NGC 281

AT THE BORDERS OF THE CONSTELLATIONS of Cassiopeia and Cepheus lies a hotbed of star formation—the Nebula NGC 281. Already, stars have been born in this glowing complex. Right at its heart is a star cluster only a few million years old. Powerful ultraviolet radiation from these energetic young stars lights up the surrounding hydrogen gas, giving it a red color. Star birth here has a long way to go: the dark bands of sooty dust hide regions where massive star formation is going on at this moment. The nebula is around 10,000 light-years away, toward the edge of our galaxy.

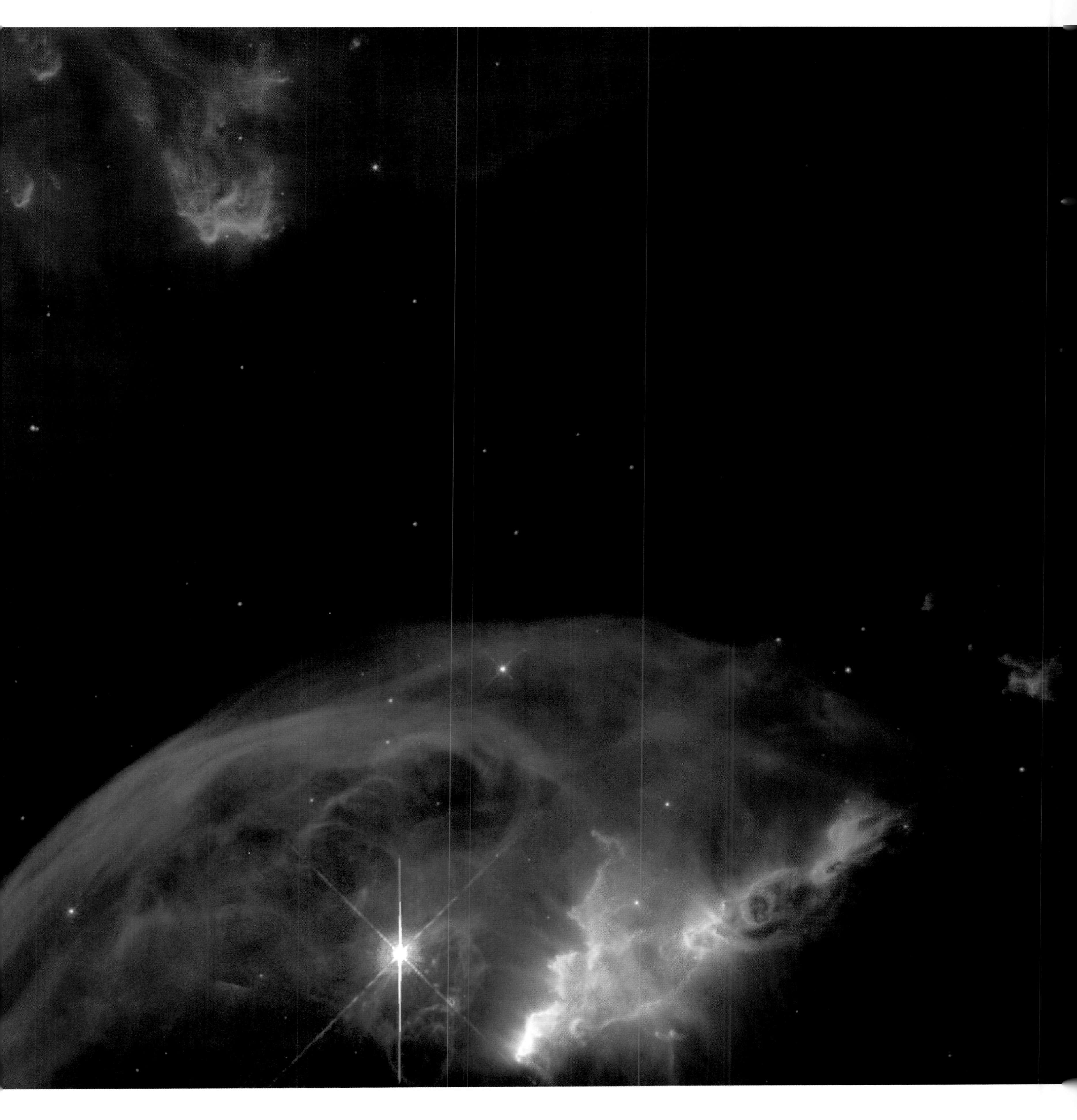

Bubble Nebula

ANOTHER SENSATIONAL SIGHT IN CASSIOPEIA, this ghostly, glowing cosmic jellyfish, is just one of three bubbles surrounding an active young star. The Bubble Nebula has been created by stellar winds—electrically charged particles pouring off in a cosmic gale from the star's turbulent surface. Some six light-years across, and expanding at a rate of 4 miles per second/7 km/s, the Bubble Nebula consists of a shell of ionized gas—in this case, hydrogen atoms stripped of their single electron. The star powering the Bubble is forty times heavier than the Sun.

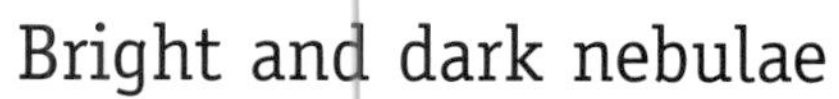

Bright and dark nebulae

THE MASSIVE STAR FIELDS TOWARD THE CENTER OF OUR MILKY WAY are packed with stellar nurseries. When you look toward the constellation of Sagittarius, in line with our galaxy's core, you see many famous nebulae—such as the Lagoon and the Trifid, which are easily visible through binoculars. Close to this pair lies the much lesser known star-forming region IC 4678. In this image of the nebula from the Canada-France-Hawaii Telescope on Mauna Kea, dark dust clouds mask glowing regions where stars have already been born.

Rosette Nebula

LOOKING LIKE A FLOWER LIGHTING UP THE DARKNESS of the cosmos, the Rosette Nebula certainly earns its romantic name. The nebula is a star-forming region some 130 light-years across and situated 3,000 light-years away, against the backdrop of the stars in Monoceros. Stellar winds from the young stars already born have blasted a central bubble in the cloud. In this image, the colors have been enhanced: red comes from the plentiful hydrogen gas in the nebula, green from oxygen, and blue from sulfur.

Flaming Star Nebula

RUNAWAY STAR: AE AURIGAE WAS BORN IN THE ORION NEBULA 2.7 million years ago. But since then, it has jetted across the sky to reside in the constellation of Auriga. No one is certain as to why a star makes a break for the border like this. One idea is that its massive companion star has exploded as a supernova—and, once it has gone, the star has nothing to hold it back. AE Aurigae has temporarily encountered a region of gas and dust. Its intense radiation is lighting up the cloud as the Flaming Star Nebula.

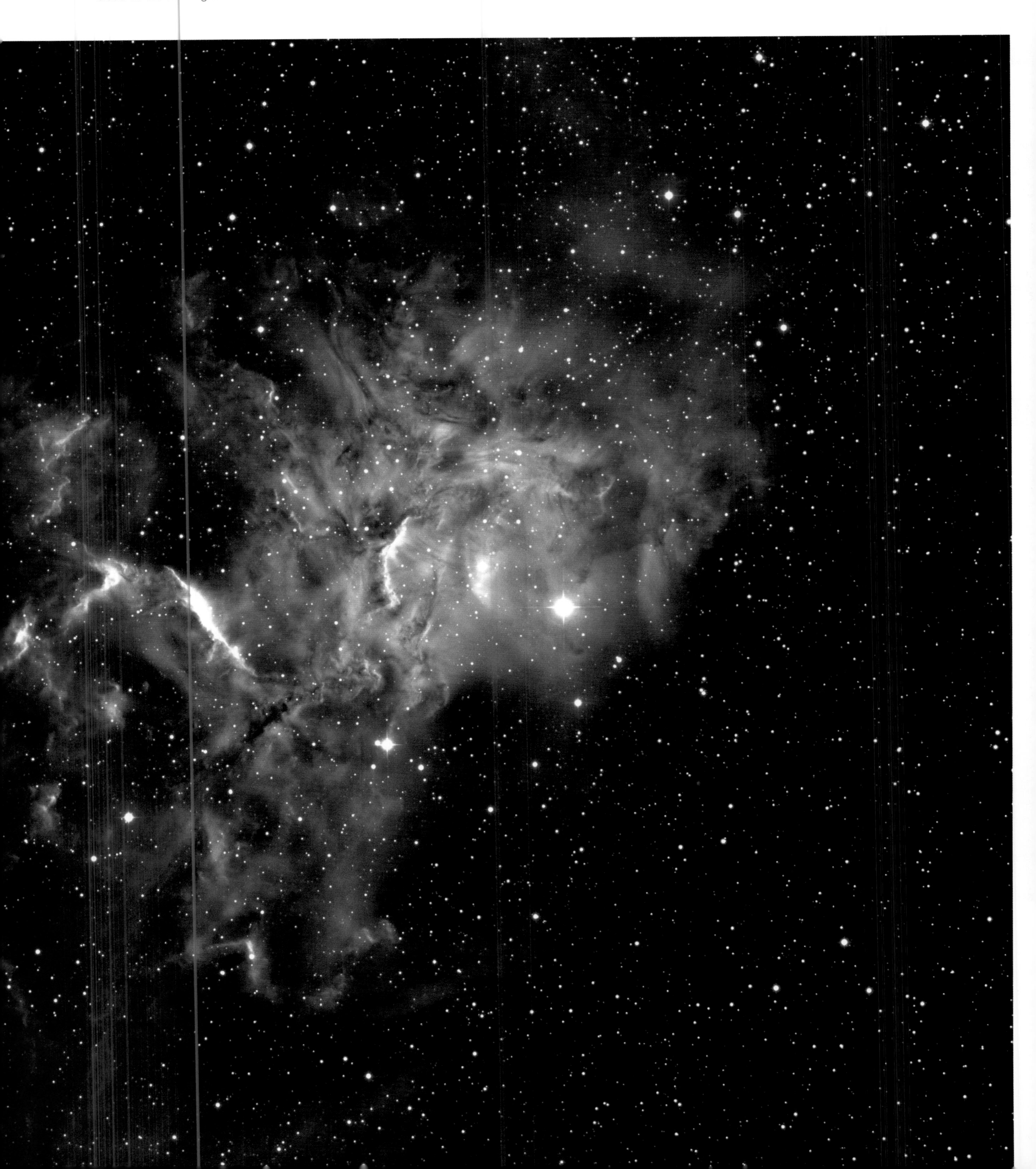

Eagle Nebula

FROM THE CLEAR SKIES OF KITT PEAK'S NATIONAL OBSERVATORY in Arizona, its 3-foot- /0.9-m-diameter telescope captured this stunning image of the Eagle Nebula in Serpens. At its center are pillars of gas—known as "elephant trunks"—photographed famously in close-up by the Hubble Space Telescope. Seen through a small telescope, the nebula looks in outline like a bird of prey, but in this detailed image, the "window" at its heart turns out to be a busy star formation workshop. Six thousand five hundred light-years away, the Eagle Nebula measures about twenty light-years across.

Omega Nebula

DISCOVERED BY THE FRENCH ASTRONOMER PHILIP DE CHESEAUX in 1746, the Omega Nebula's name derives from its shape. De Cheseaux described it as having "the perfect form of a ray . . . its sides are exactly parallel." Later, large telescopes revealed a curve of gas springing from one end of the bar, making it look like the Greek letter omega. The Omega Nebula—pregnant with gas—is illuminated by stars hidden within the neighboring dark clouds.

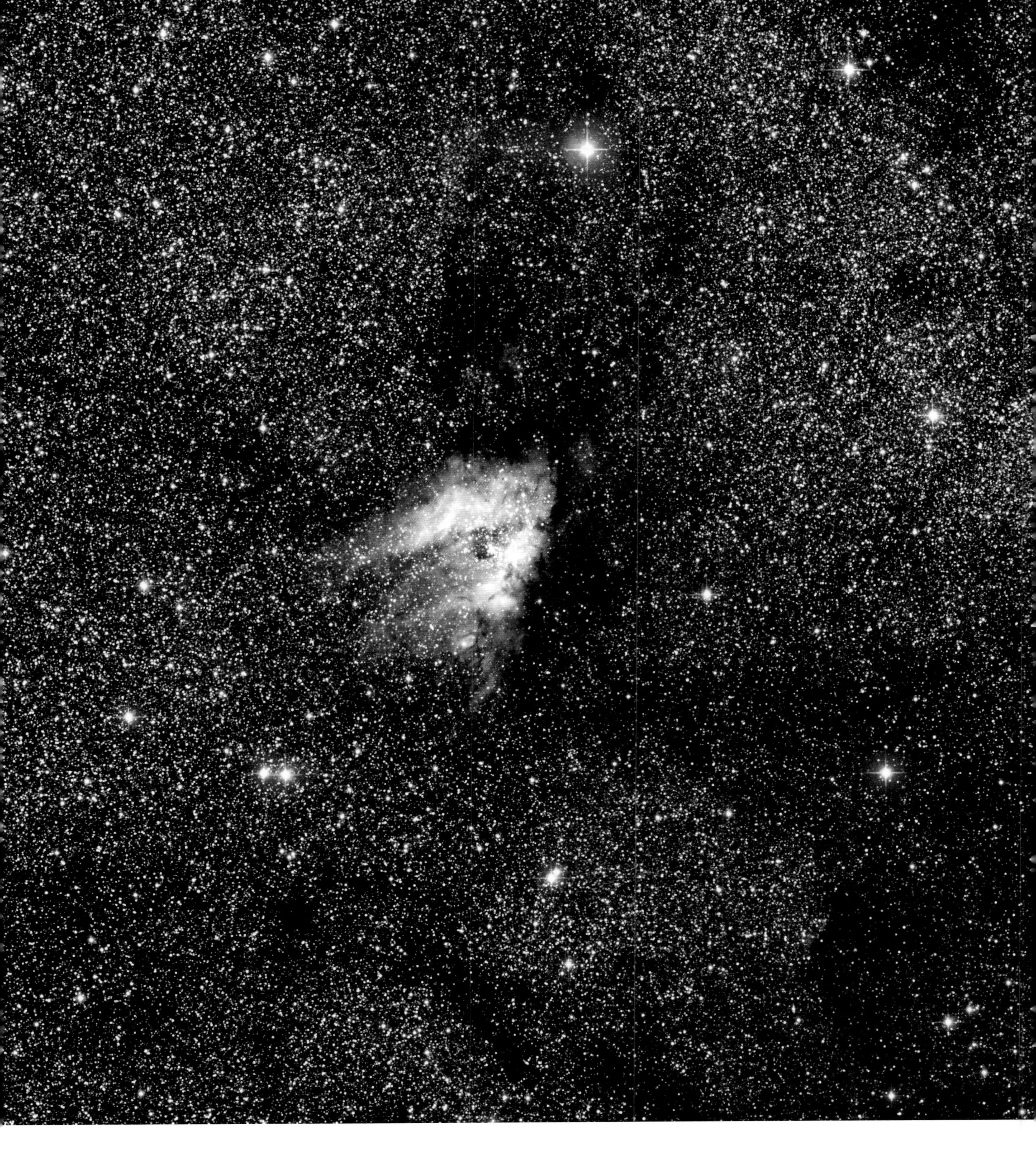

Star birth in the Omega Nebula

THIS WIDER-ANGLE VIEW OF THE OMEGA NEBULA reveals the surrounding dark dust and gas cloud that led to the nebula's formation and conceals some very hot young stars. The heated hydrogen gas blazes out in the center of the frame. Eventually, all of the dark wispy fronds will light up in a fury of star birth: they contain enough material to give birth to a million stars. The myriad background stars are other suns: many of them will have their own planets, possibly with life.

Cone Nebula ▷

BOTH PART OF A HUGE STAR-BIRTH COMPLEX IN MONOCEROS, the Fox Fur Nebula and the Cone Nebula (seen here) are stellar maternity wards. The Cone is seven light-years long, but in this image from the latest camera on the Hubble Space Telescope, we only see the top two-and-a-half light-years. It is being sculpted by powerful winds and radiation from a star that lies off the top of this image. But this process is also sweeping away the natal gas from any stars that may be forming within. So these fledgling stars, in the Cone's dark interior, face a race against time if they are to be born.

Cone and Fox Fur Nebula

THE BEAUTIFUL NEBULA AT THE LOWER RIGHT OF THIS IMAGE is nicknamed the Fox Fur Nebula after its resemblance to a 1930s fox-fur evening stole. Its texture is derived from the complex interactions between young stars and cosmic dust: they drive the sooty matter into tapestrylike convolutions. The dust is also reflecting the light from some of the brightest newly formed stars, giving the nebula an eerie blue glow. On the far left, the Cone Nebula's dark head intrudes into the redness of the energized and heated hydrogen gas.

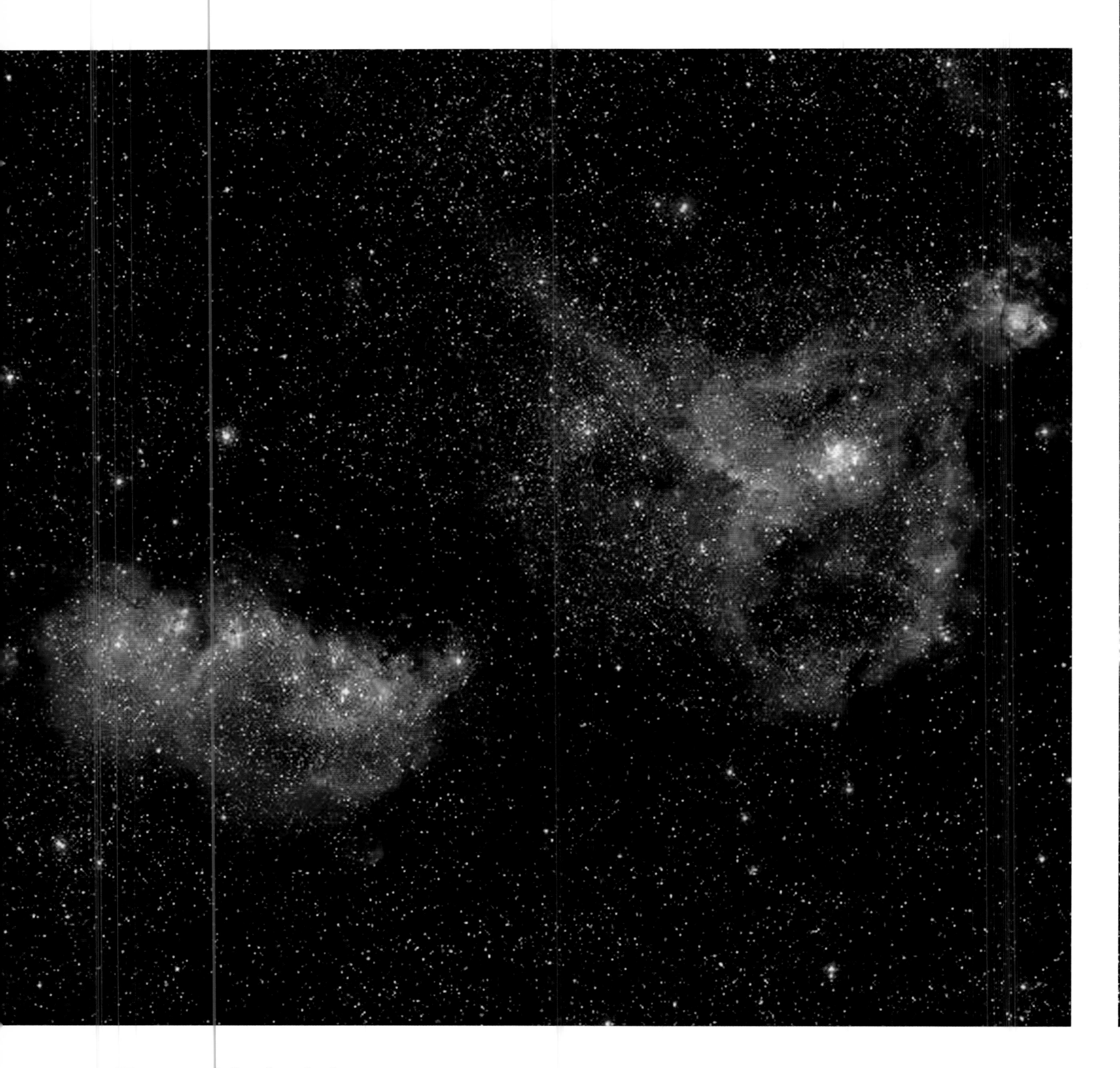

Heart and Soul nebulae

DO THE HEART AND SOUL OF OUR MILKY WAY LIE IN CASSIOPEIA? The Heart and Soul nebulae certainly do. The Heart (right), is named after its resemblance to our body's most vital organ; the Soul (left) is more ghostly and unstructured. These nebulae are the silver lining of a huge dark cloud 6,000 light-years away, in which stars have already begun to form. This is a vast complex: the view here extends four Moon-widths, and it is creating the most massive and brightest stars in the galaxy. The hottest area of action is the knot at top right of the Heart Nebula, where an enormous outburst of star formation is taking place.

Tarantula Nebula

THIS COSMIC ARACHNID HAS ITS LAIR IN OUR NEAREST NEIGHBOR-GALAXY, the Large Magellanic Cloud. One hundred sixty thousand light-years away, the Tarantula Nebula—which is 1,000 light-years in diameter—is one of the biggest known nebulae. Its central star cluster energizes this vast cloud of gas, lighting up the spidery tendrils. These stars are immensely hot, with temperatures of 54,000 degrees Fahrenheit/30,000 degrees Celsius (as compared to around 11,000 degrees Fahrenheit/6,000 degrees Celsius for our Sun). They are also massive, weighing in at 60–100 Suns. This image was captured by the 7-foot- /2.2-m-telescope at the European Southern Observatory in Chile.

Iris Nebula

LOOKING LIKE A MASS OF CELESTIAL PETALS, the Iris Nebula is the only celestial object to be named after a flower. The beautiful blue color of the nebula is due to the scattering of light by dust—the same process that makes our sky blue. This is not a star-forming region but a "reflection nebula" created by light from the young star inside the gaseous shroud bouncing off the dust particles. The Iris Nebula is six light-years across and lies about 1,300 light-years away in the constellation Cepheus.

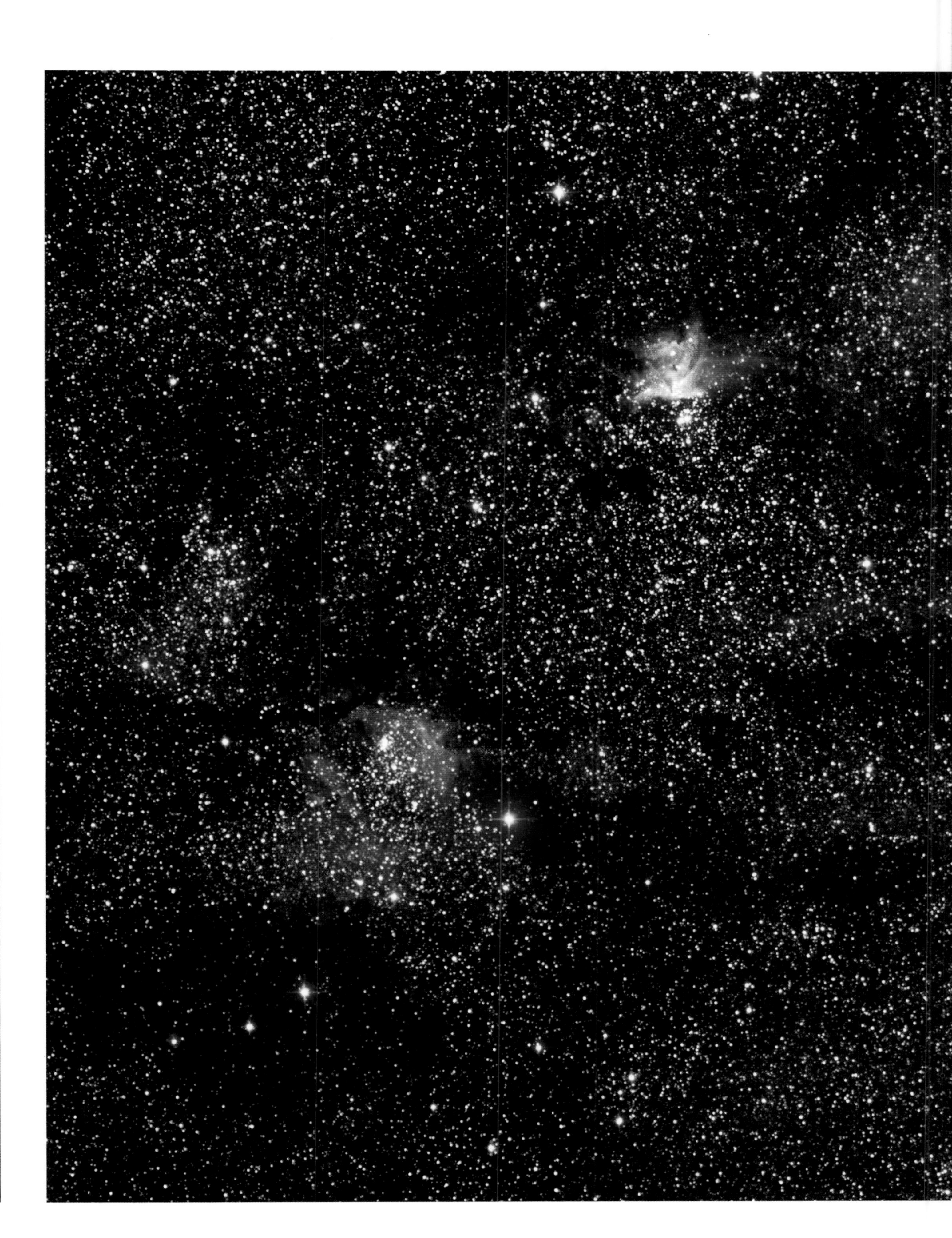

War and Peace Nebula

FIVE-AND-A-HALF THOUSAND LIGHT-YEARS AWAY in the constellation of Scorpius, the War and Peace Nebula blazes out against a background of distant stars. It derives its name from the shape of the gas clouds. On the top right is the brilliant "Peace," which looks like a flying dove. Below left is "War." Although not obvious in this particular image, the dark star-forming clouds (center) can look like the sunken eyes of a skull, while the upper blue region represents its forehead and the lower area its chin.

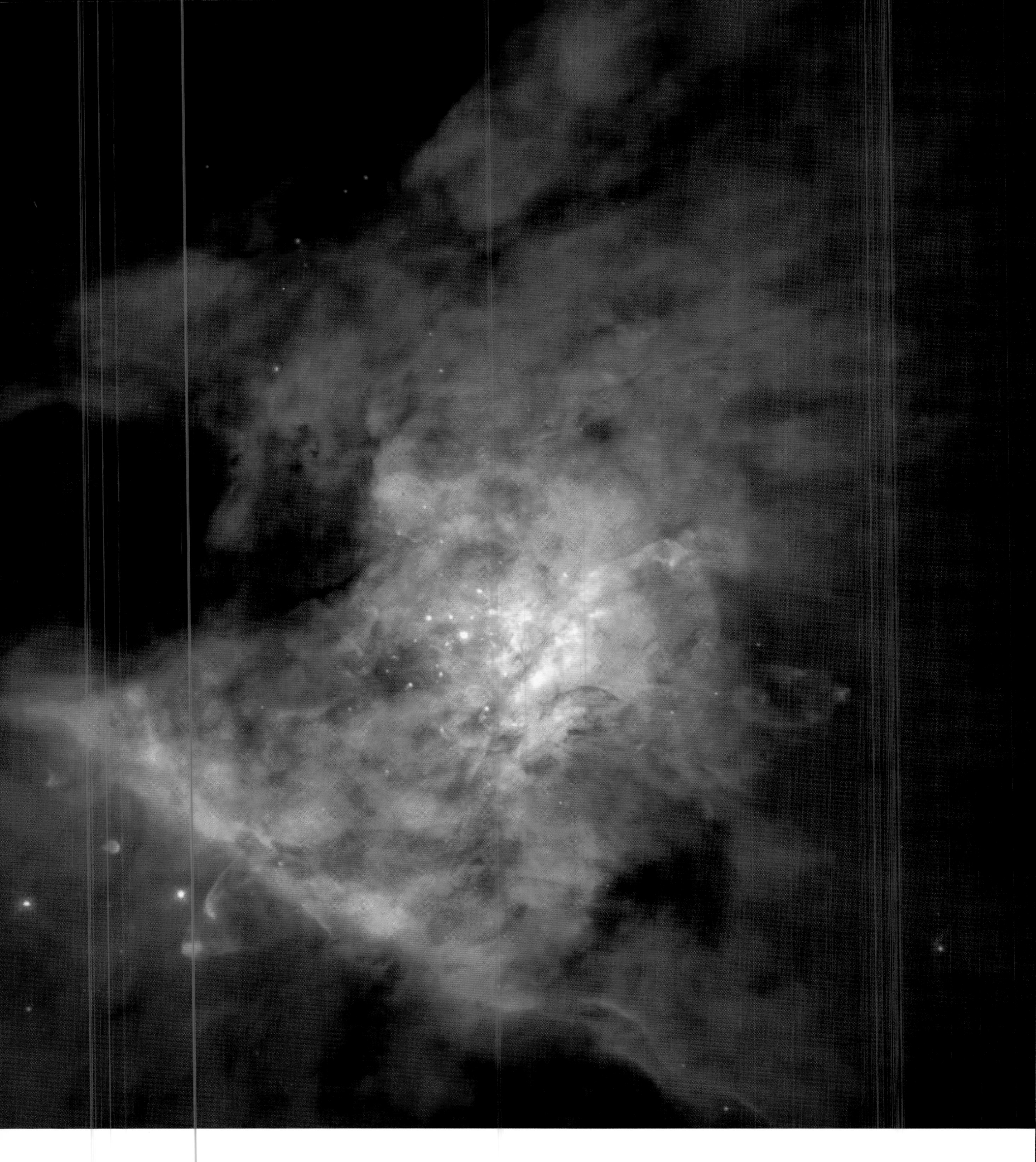

Orion Nebula

THE MOST FAMOUS GAS CLOUD IN THE SKY, the Orion Nebula is the center of the "sword" hanging below Orion's three-starred belt. It is easily visible to the unaided eye as a fuzzy patch. Through a small telescope, it is a spectacular sight: a gas cloud lit up by a quartet of newly born stars known as the "Trapezium" (center). These energetic stars illuminate the nebula, while their ferocious stellar winds and pressure from radiation cause vast structures to form in the surrounding dust and gas. The "Great Wall" (the line below the Trapezium) is a massive shock wave.

Orion Nebula

FIFTEEN LIGHT-YEARS ACROSS, THE ORION NEBULA is just a tiny fragment of a vast star-forming region that covers the whole of the constellation. Almost all of the stars in Orion are young and were born in the same star cloud. This is a rarity in the cosmos: most stars making up constellations lie at different distances. Many stars still have yet to be born. This image—amazingly, taken with an amateur telescope—sums up the incredible level of cosmic turmoil in the region. Above the Orion Nebula, the Nebula M43 glows blue from starlight reflected off dust particles.

Horsehead Nebula

STUNNINGLY SILHOUETTED against the warmer, red regions of the Orion star-forming region, the Horsehead stands proud above the giant dark cloud that spawned it. The cloud (in the lower part of the image) has been eroded by light from stars nearby. It has also been compressed by an expanding shell of hydrogen gas some fifty light-years across—one of many in the Orion region. However, the Horsehead is made of such dense gas that it has been able to survive. Like clouds on Earth, its shape will change over the millennia.

Previous pages | Hubble's sharpest view

IN ONE OF THE MOST DETAILED ASTRONOMICAL IMAGES EVER PRODUCED, NASA's Hubble Space Telescope captured an unprecedented look at the Orion Nebula. This turbulent star formation region is one of astronomy's most dramatic and photogenic celestial objects. "Orion is a bustling cauldron of activity. This new large-scale Hubble image of the region reveals a treasure-house of beauty and astonishing detail for comprehensive scientific study," said Jennifer Wiseman, NASA's Hubble program scientist. The image reveals 3,000 previously unidentified newborn stars.

Eta Carinae

A GAS CLOUD POISED TO CREATE A MILLION STARS, the Carina Nebula is a staggering 9,000 light-years away, and yet it appears twice as large as the full moon. It cloaks one of the brightest and most massive stars in the galaxy: Eta Carinae. This is one of the most wayward objects in our Milky Way. In 1837, it was a very average star, but six years later, it became—temporarily—the second brightest star in the sky, outshining our Sun five million times over. Although it is a young star, it is so unstable that it will almost certainly explode within the next few thousand years.

Lagoon Nebula

THE LAGOON NEBULA (CENTER)—named for the lagoonlike dark patch that crosses its center—lies close to a cluster of other nebulae, creating the impression of a cosmic string of pearls. Five thousand light-years away in the direction of Sagittarius, the Lagoon is a classic star factory. It boasts a young cluster of stars that is now largely free from their natal gases, newly hatched stars just emerging from the cocoon, and protostars still hidden in the dense dark cloud that surrounds the whole nebula.

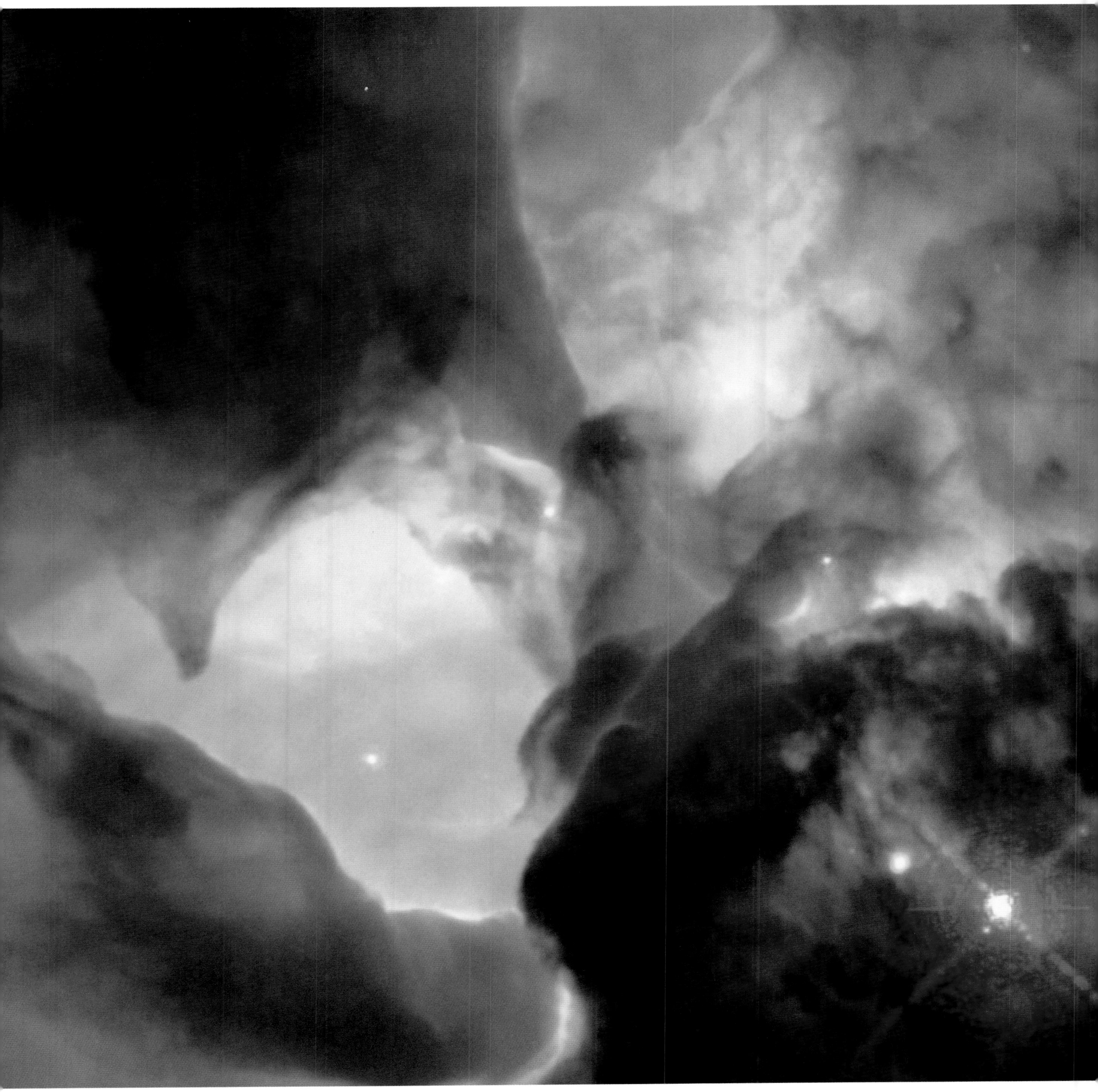

Tornado structures in Lagoon Nebula

THIS HUBBLE SPACE TELESCOPE IMAGE REVEALS A PAIR OF INTERSTELLAR "TWISTERS," both measuring half a light-year—eerie funnels and twisted rope structures—in the heart of the Lagoon Nebula. The central hot star, O Herschel 36 (lower right), lights up the brightest region in the nebula, called the Hourglass. Other hot stars also create violent stellar winds, which tear into the cool clouds. Like the spectacular phenomenon of tornadoes on Earth, the large difference in temperature between the hot surfaces of the stars and cold interior of the clouds—combined with the pressure of starlight—may produce a strong horizontal shear to twist the clouds into their tornadolike appearance.

Trifid Nebula

ITS NAME HAS NOTHING TO WITH JOHN WYNDHAM'S FAMOUS SCIENCE FICTION NOVEL *The Day of the Triffids*; instead, this nebula is divided into three by dark lanes of cosmic soot. John Herschel—son of William, discoverer of the planet Uranus—described it in the nineteenth century as "singularly trifid, consisting of three bright and irregularly-formed nebulous masses." It lies close to the Lagoon Nebula in Sagittarius. This dramatic image from the Spitzer Space Telescope reveals the heat radiation coming from the nebula. It penetrates the dusty veils of the Trifid and exposes the young stars being born within.

NGC 1333

THE SPITZER SPACE TELESCOPE ALSO CAPTURED THIS IMAGE of the Nebula NGC 1333 in the constellation of Perseus. Just 1,000 light-years away—a short distance on the scale of the universe—stars have recently been born here. In their youthful enthusiasm, the stars are firing powerful jets of gas into the nebula that once gave birth to them. Researchers have never before seen so many stellar jets in one single location. It is highly possible that this violence may disperse and destroy the nebula—which will prevent further star birth here forever.

STARS HAVE A PROBLEM WHEN THEY RUN OUT OF FUEL: THEY DIE. And it is ironic that some of the most beautiful sights in the universe are the result of star death.

Like our Sun, stars live as a result of nuclear fusion reactions. Deep in their cores—where the temperature runs into millions of degrees—the heat and pressure is so aggressive that the star fuses its hydrogen into helium. It is an energy source that keeps the star shining—but only for so long.

For an average star like the Sun, "so long" is around ten billion years. At 4.6 billion years of age, our star is about halfway though its life. But it will eventually use up the hydrogen in its core. Bereft of a power source, the core will shrink—and, counterintuitively, it will grow hotter.

This will cause the outer layers of the Sun to billow out. Cooling as the layers expand, the color of our star will redden. Five billion years hence, our Sun will have become a red giant. Its growing girth will swallow up the innermost planets Mercury and Venus and parch the oceans from the Earth. There will be no future for humankind.

Eventually, the Sun will become so unstable that it will start to shed its atmosphere into space as a glorious, glowing "Planetary Nebula." These lovely, short-lived cosmic apparitions take on the most amazing shapes, from celestial butterflies to hourglasses.

So what is left? All that the Sun will leave behind is its cooling core: a sad little star called a white dwarf, which will eventually become a dead, black cinder.

When it comes to stars heavier than the Sun, there is another option—but there is still no cheating death. Because they are so massive, these stars have the gravity to consume their fuel supplies at an obscene rate. When one particular fuel—such as hydrogen—is exhausted, they tap into the next, which is helium. When the helium runs out, they start to fuse its end-product, carbon. And so on.

The end is nigh when the star starts to fuse iron. This reaction is hugely unstable. The star has to take *in* energy. It responds by contracting—and the resulting *implosion* results in a spectacular *explosion*: a supernova.

These cosmic suicides are among the most incredible—and harrowing—sights in the universe. A supernova can rival the brightness of its galaxy, which contains billions of stars. However, it also leaves behind a beautiful reminder of its life: a glowing supernova remnant, which may last for 100,000 years.

The core of a supernova is too massive to become a mere white dwarf. It may turn into a neutron star or a black hole, but more of that later.

At the end of the day, dying stars bequeath so much to the universe. Because they have spent their entire lives processing complex elements—and then ejecting them into space—they sow the seeds of life into the cosmic melting pot. From the carbon in your bones to the gold in your wedding ring, you must thank dying stars: their deaths made it all possible.

6 STAR DEATH

Dumbbell Nebula

THE DUMBBELL IS A PLANETARY NEBULA—so-called because, in the late eighteenth century, the renowned astronomer William Herschel (who discovered Uranus) compared them to planetary disks. But these are actually stars on the way out, ejecting material from their bloated, aging bodies. The Dumbbell is bipolar in shape: the material has been flung out into space from opposite sides of the star. In this image, the ejecta is color coded—hydrogen is shown as blue, sulfur is green, and helium is red.

Previous pages | **Helix Nebula**

THIS CLOSE-UP IMAGE OF THE HELIX NEBULA, captured by the Hubble Space Telescope, reveals matter from a star ejected in its death throes. But a phoenix will arise from the ashes. The cast-off gas is rich in newly created elements, such as carbon and oxygen, providing the raw materials for a new generation of stars, planets, and life.

Egg Nebula

ANOTHER STAR AT THE END OF THE ROAD: the Egg Nebula (sometimes called the Rotten Egg Nebula) is a planetary nebula in which the dying star is ejecting matter at a rate of around 621,000 miles per hour/1 million kph. In this Hubble Space Telescope image, the central star is obscured by a dark pall of dust. This detailed picture reveals how stars jettison their gas: in the case of the Egg Nebula, it streams out in beams that look like cosmic searchlights.

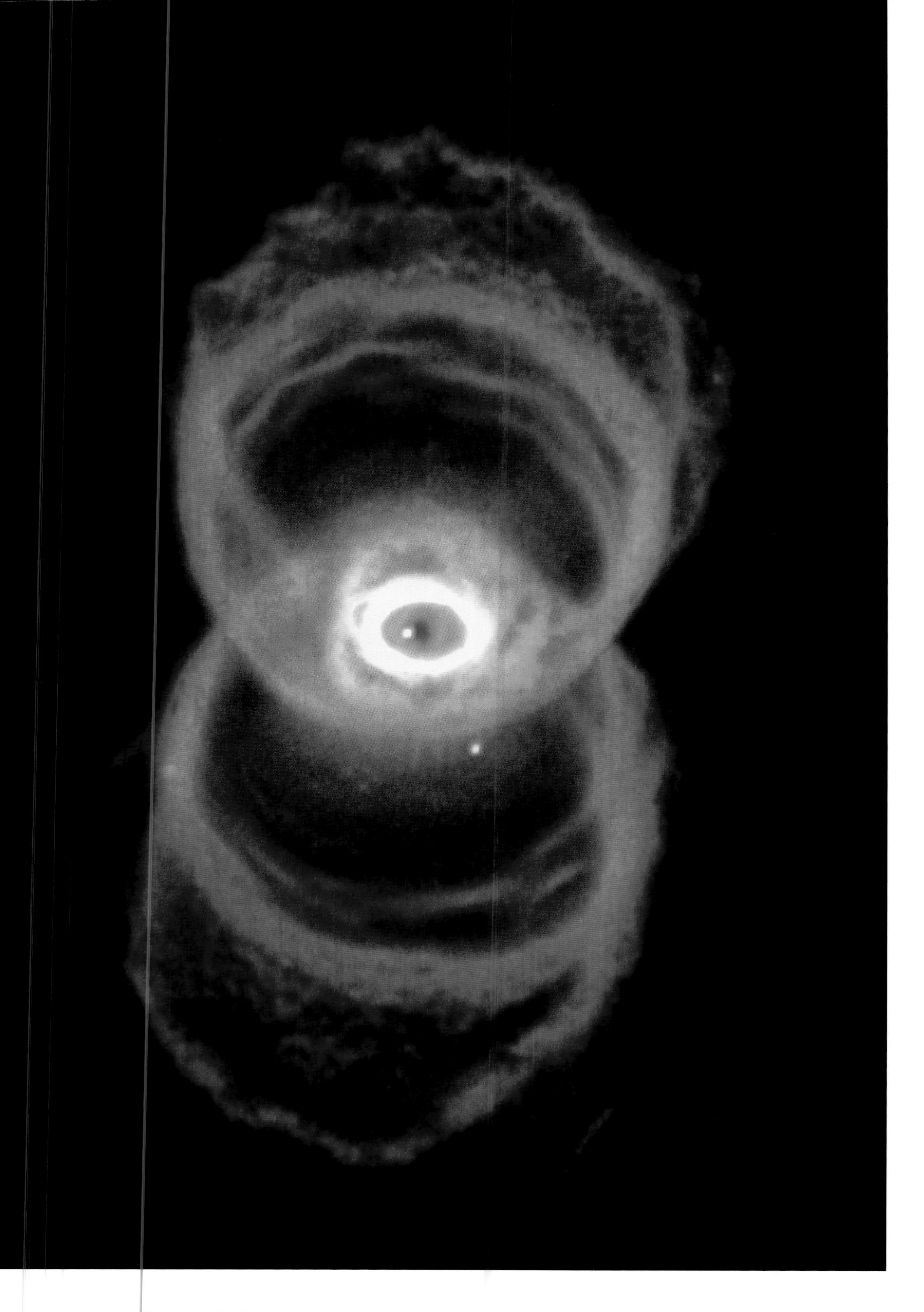

Hourglass Nebula

A HUBBLE SPACE TELESCOPE IMAGE of the evocatively named Hourglass Nebula. Some 8,000 light-years away, this is another star for whom the sands of time are running out. Astronomers believe its hourglass shape derives from massive outflows of gas encountering previously ejected material that is denser at the equator than at the poles. At the center is the (white) core of the former star. It is slightly displaced—perhaps indicating that it is being pulled on by an unseen companion star.

Cat's Eye Nebula

THREE THOUSAND LIGHT-YEARS AWAY, the Cat's Eye Nebula is one of the most complex planetaries known. Like the Hourglass, the central star is almost certainly double (although its partner is not visible). Only this could account for the convoluted arcs and shells blown off by the main star, caused by the companion tugging its more massive partner off course. The nebula is estimated to be around 1,000 years old, and—in the words of its researchers—represents the "fossil record" of a dying star.

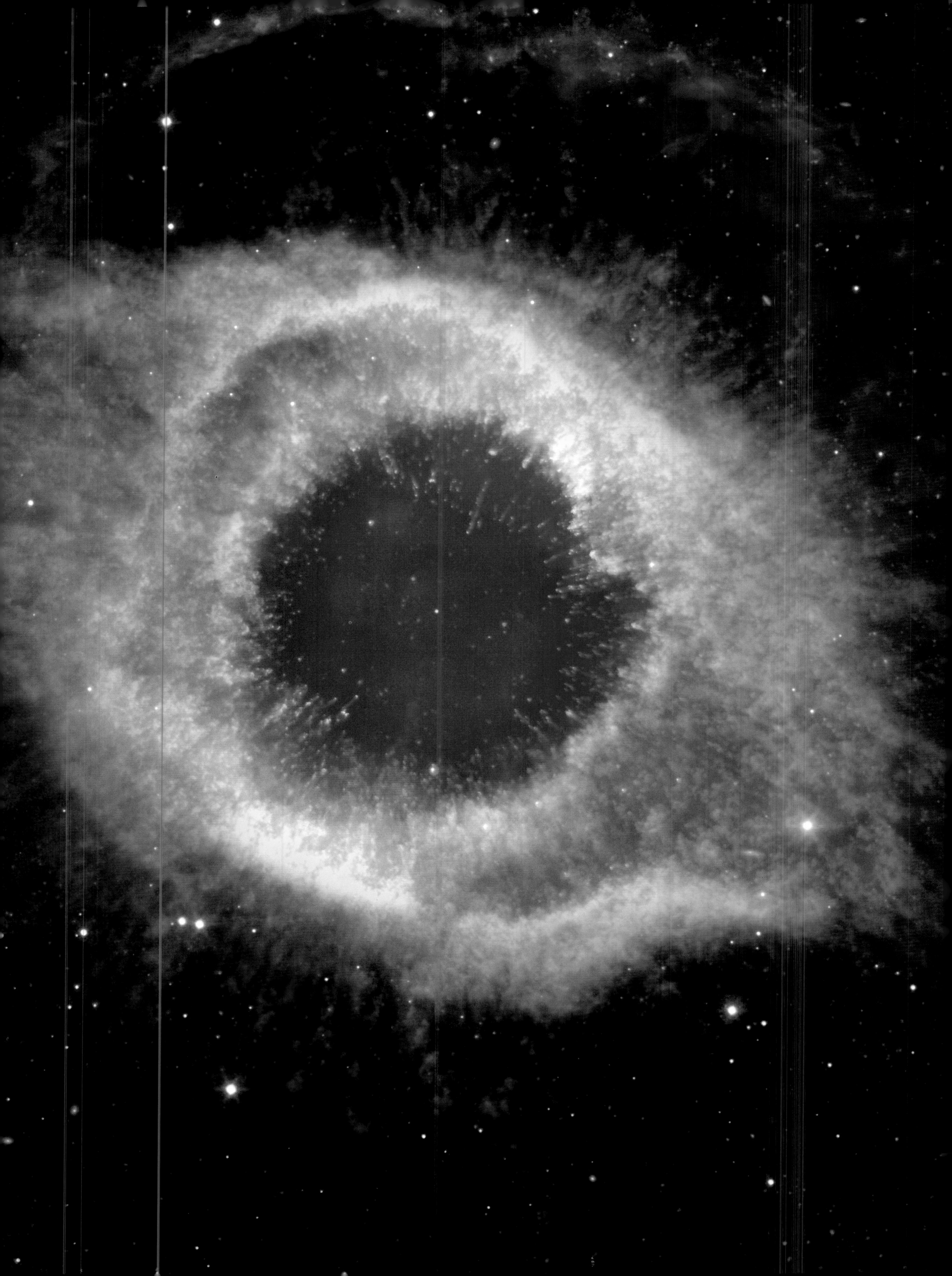

◁ Helix Nebula

THIS BRAND-NEW IMAGE OF THE HELIX NEBULA was achieved by combining data from the Spitzer orbiting telescope and the Hubble Space Telescope. We are looking directly down the helix in this three-dimensional image. Approximately 450 light-years away, the Helix is the nearest planetary nebula to the Earth, and it appears as the largest in the sky. The expanding nebula measures 1.5 light-years across. At its center lies the corpse of the original star, which is so hot that it is causing the gas in the Helix to fluoresce.

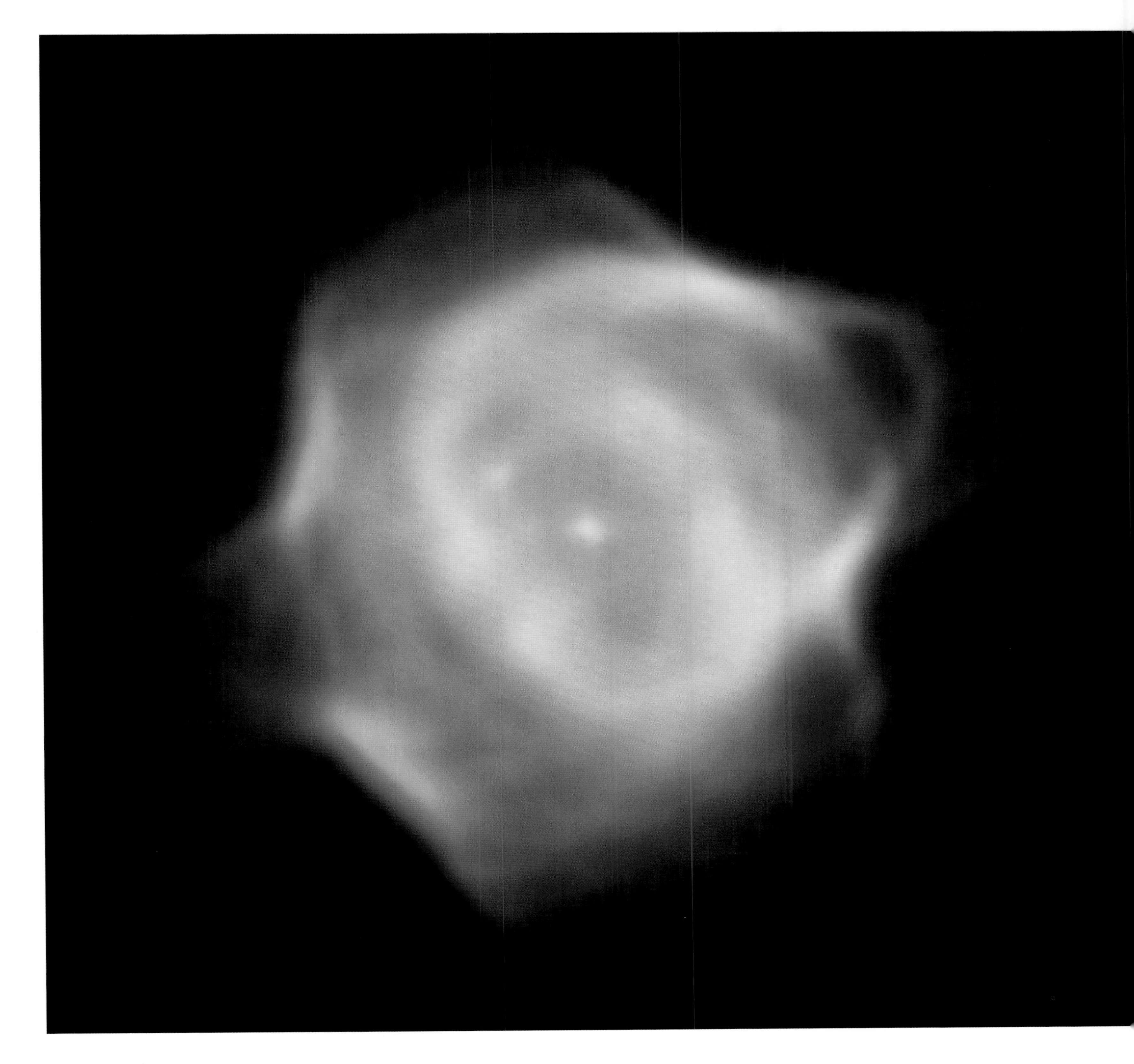

Stingray Nebula

A NEW KID ON THE BLOCK: the Stingray Nebula. Named after its graceful, gentle curves, the nebula did not exist in the 1970s. But this Hubble image, captured in March 1996, shows an expanding shell of gas around the central star. This makes the Stingray the youngest planetary nebula known. In the ten o'clock position is a companion star, whose gravity contributes to the contortions and convolutions in the bubble. The nebula is 130 times the diameter of our solar system and lies 18,000 light-years away.

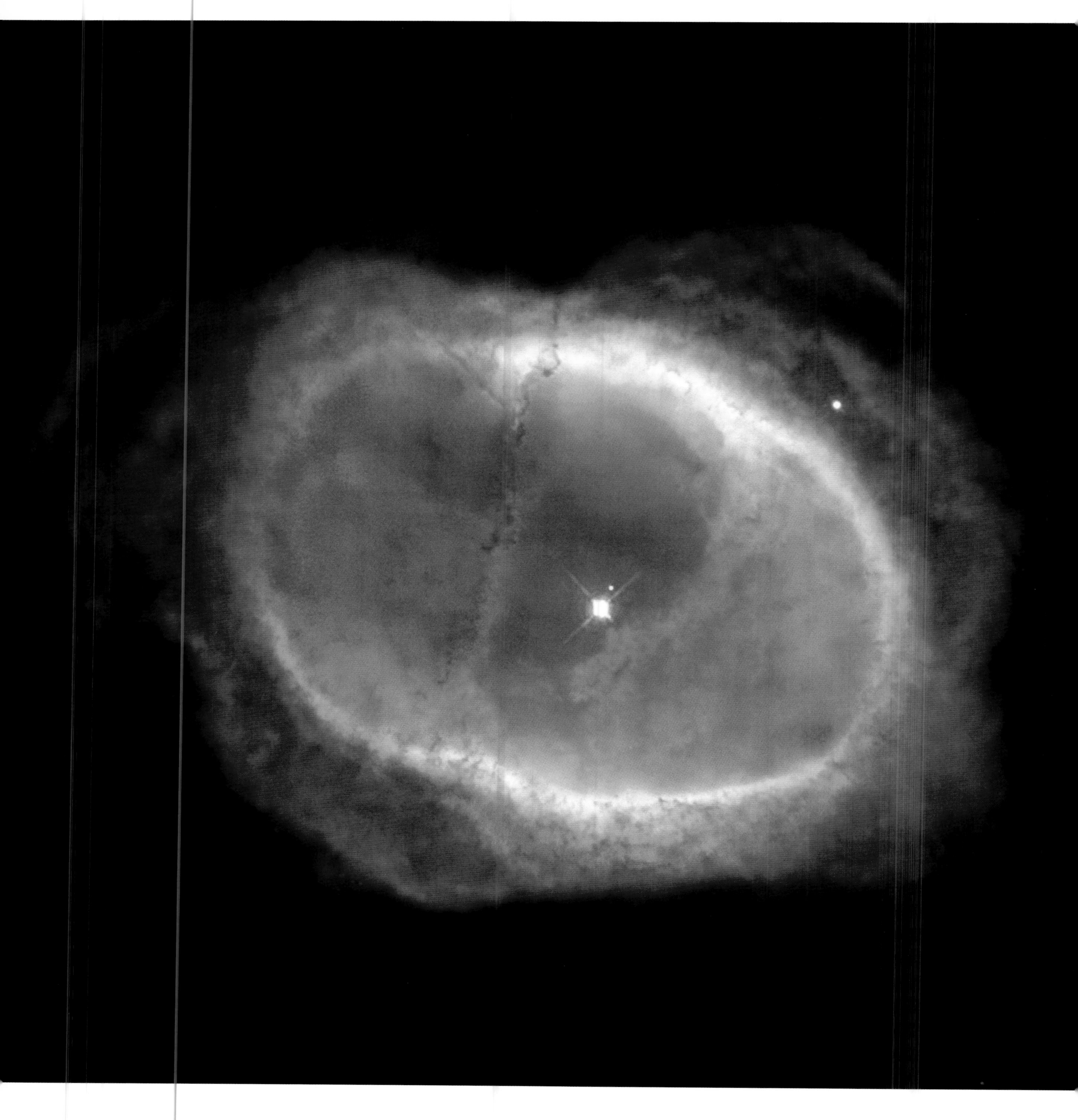

NGC 3132

NICKNAMED THE EIGHT-BURST NEBULA AFTER ITS SHAPE, Planetary Nebula NGC 3132, lies in the southern hemisphere constellation of Vela. It is also known as the Southern Ring Nebula because of its uncanny resemblance to the Northern Ring Nebula in Lyra. Around 2,000 light-years away, the Eight-burst is quite small. Blue areas are made of hot gas, while the brown and yellow outskirts are cooler material. The nebula is unusual in being crisscrossed by avenues of cool, sooty dust. All the material has been ejected by the fainter star at the center of the image—not the bright one.

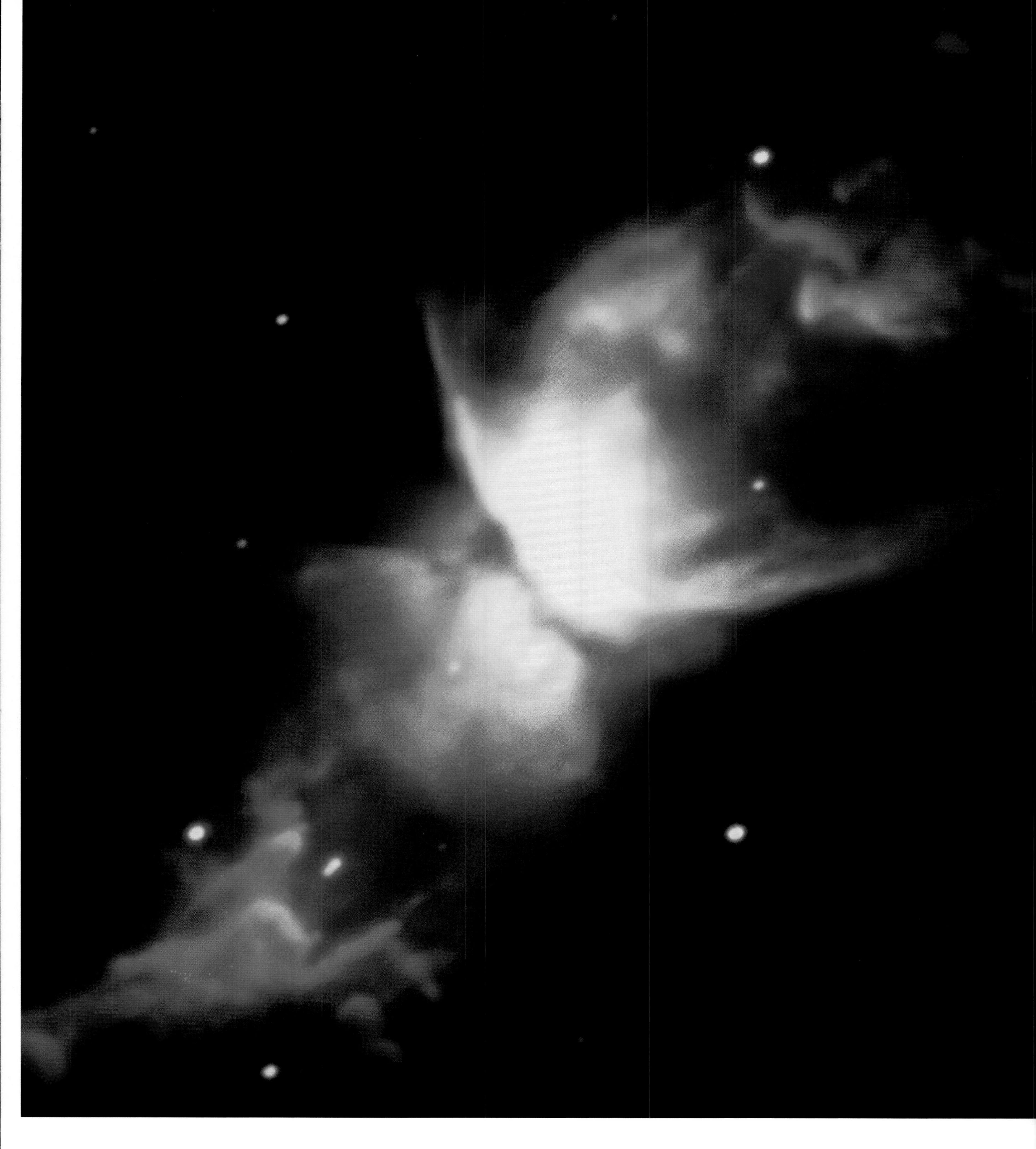

Butterfly Nebula

ASTRONOMER BRUCE BALICK, OF THE UNIVERSITY OF WASHINGTON IN SEATTLE, once memorably asked if stars are better remembered for their art after they die? Never has this been so true as in the case of the Butterfly Nebula, an aptly named planetary just over 2,000 light-years away. There undoubtedly is beauty in stellar death. This image—one of the first captured by the 27-foot-/8.2-m-diameter "Antu" instrument at the Very Large Telescope complex in Chile—shows glowing gas colored by the energy of the dying star. As in so many planetary nebulae, the central star is double.

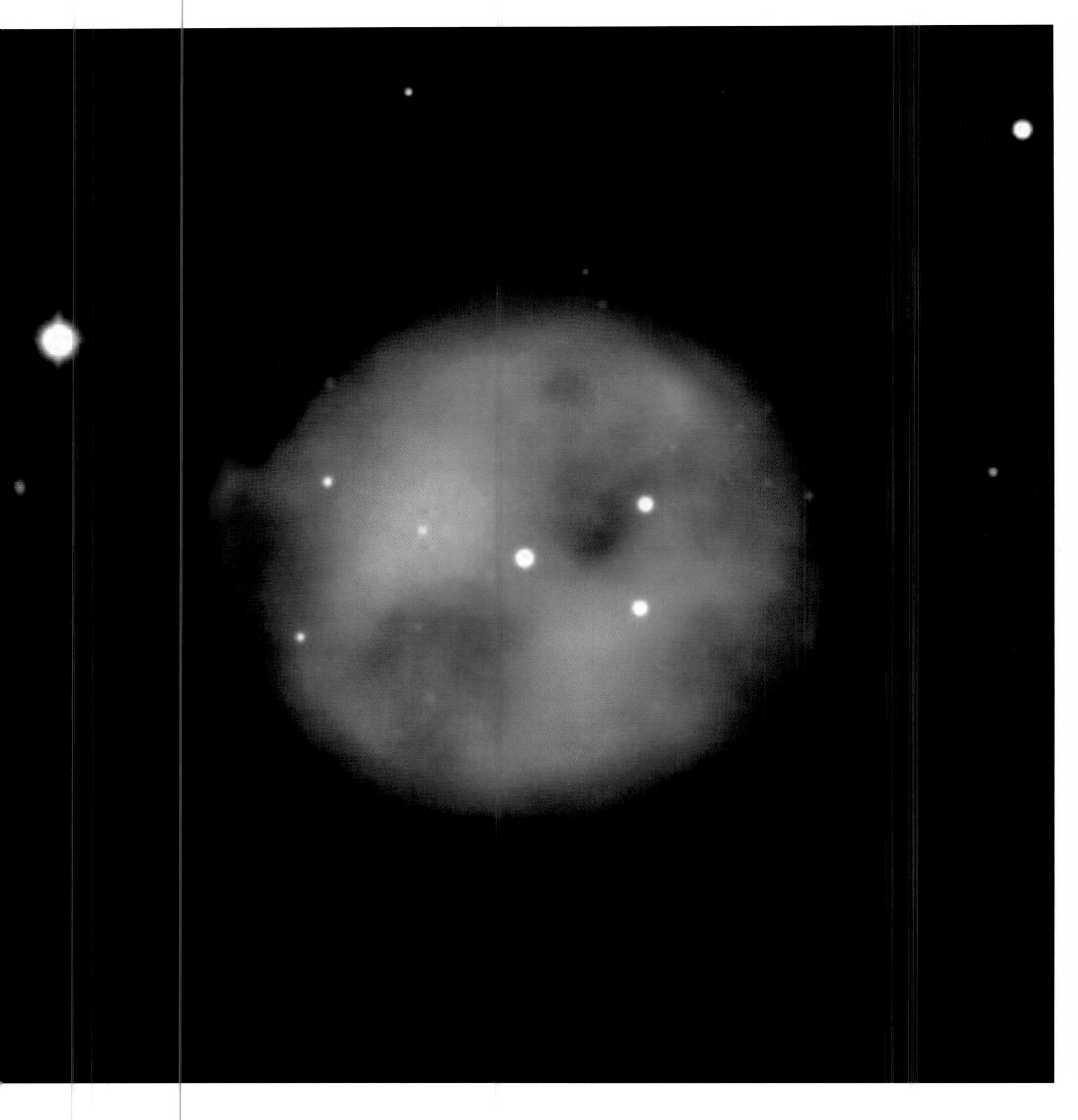

Owl Nebula

DISCOVERED IN 1781 BY PIERRE MECHAIN, the Owl Nebula was given its name in 1848 by the pioneering Irish astronomer and aristocrat Lord Rosse, who once owned the biggest telescope in the world. An observing companion, Dr. T. R. Robinson, described this planetary as "looking like the visage of a monkey." Lying in the constellation of Ursa Major, the Owl is over 2,500 light-years away and is estimated to be around 6,000 years old. Its "eyes" are regions in the three ejected shells where the gas is less dense.

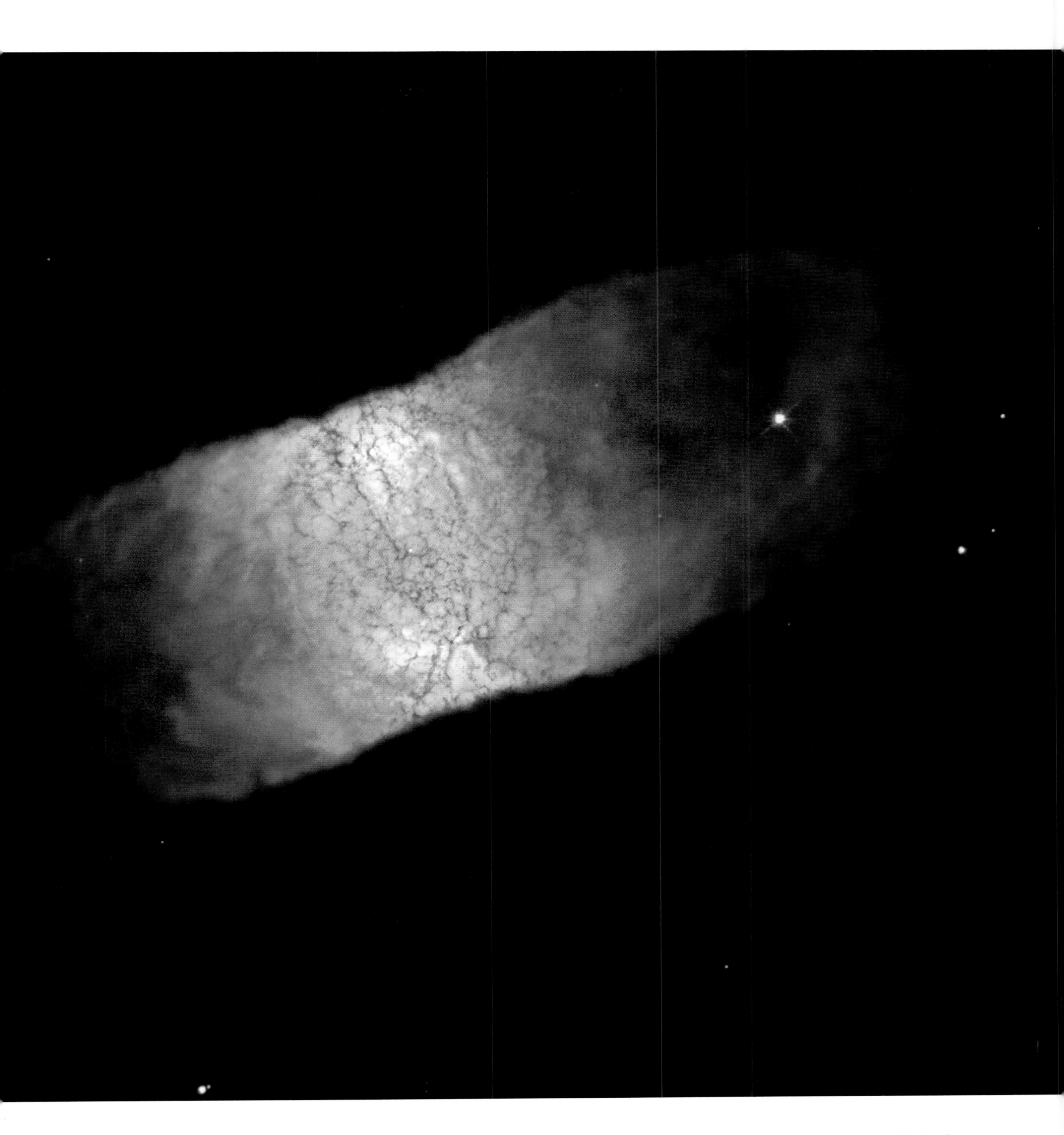

Retina Nebula

HOW CAN DEBRIS EXPELLED FROM A ROUND STAR LOOK ALMOST SQUARE? It can appear so if you are viewing it from the edge as is the case here. If you were able to circle the Retina Planetary Nebula in a spacecraft, you would discover that it is actually donut shaped. Located 3,500 light-years away, it gets its name from the intricate filaments at its center, which resemble the patterns in the retina of the eye. This image was captured by the Hubble Space Telescope.

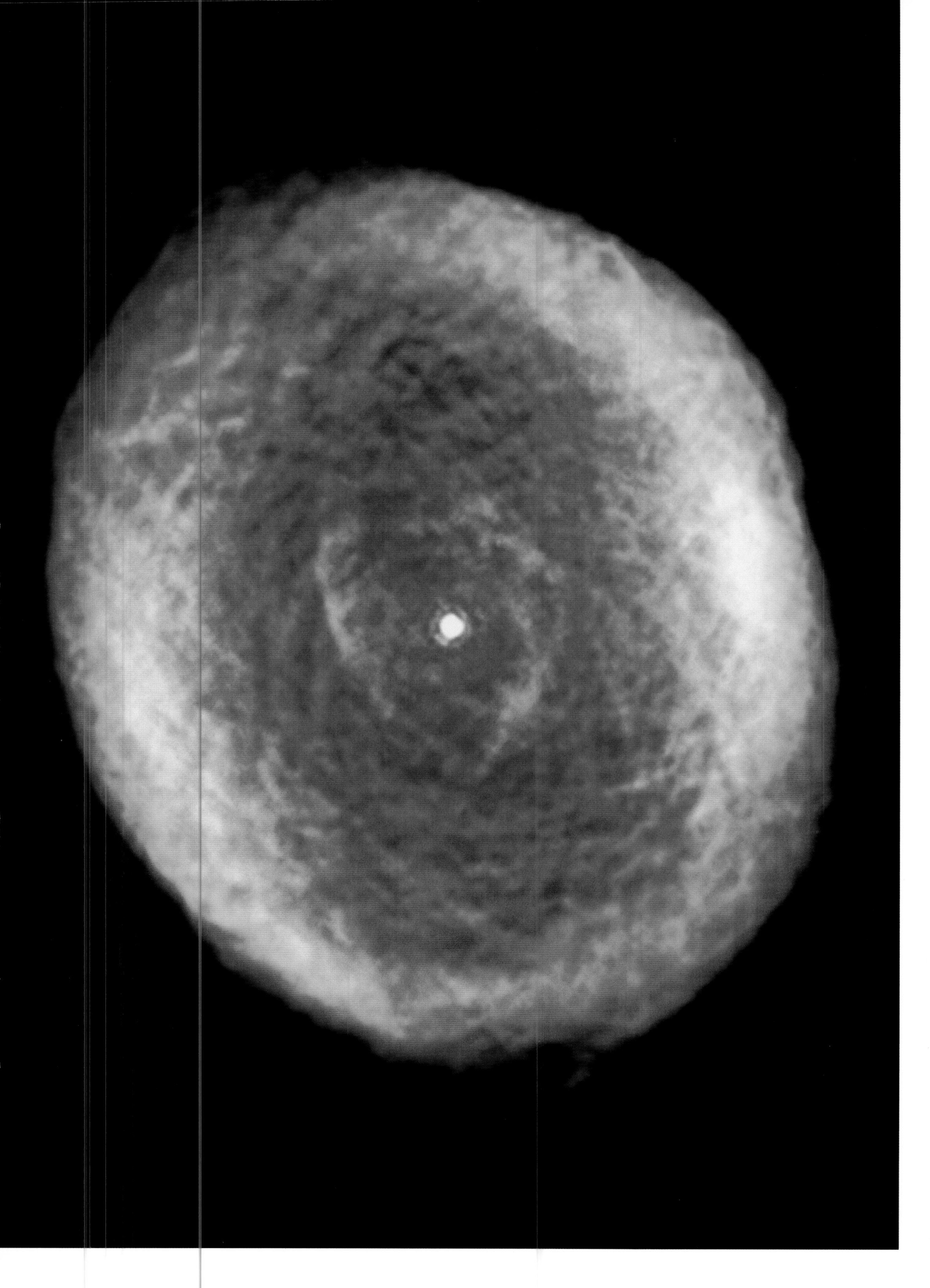

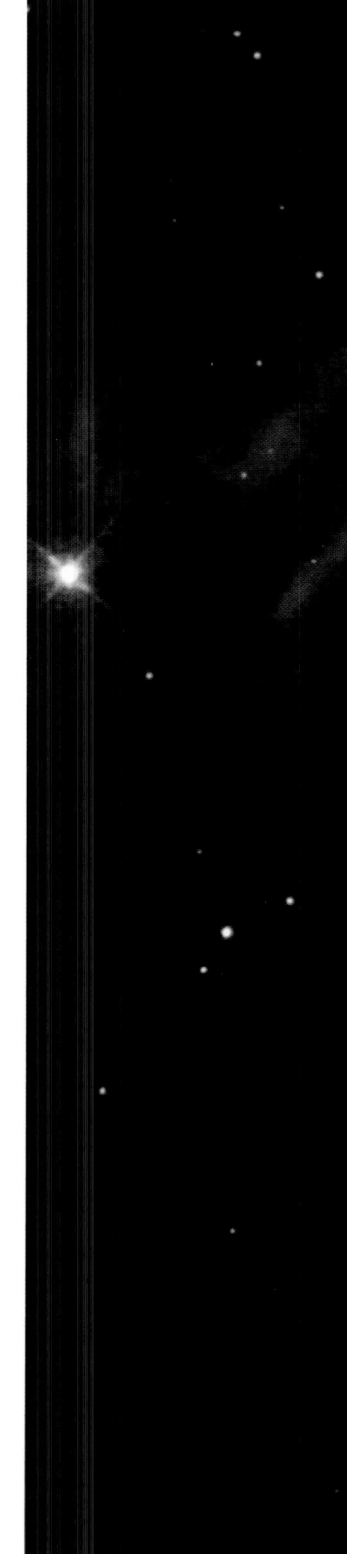

Planetary Nebula IC 418

THE APTLY NAMED SPIROGRAPH PLANETARY NEBULA glows like a multifaceted jewel against the blackness of space. It has the most baffling structure of all, making astronomers hard-pressed to come up with an explanation for its filigreelike appearance. The best bet is that chaotic stellar winds boiling off the dying star are the cause. This false-color image from the Hubble Space Telescope shows oxygen gas (hot and blue), hydrogen (cooler and green), and, furthest out, nitrogen (coolest and red).

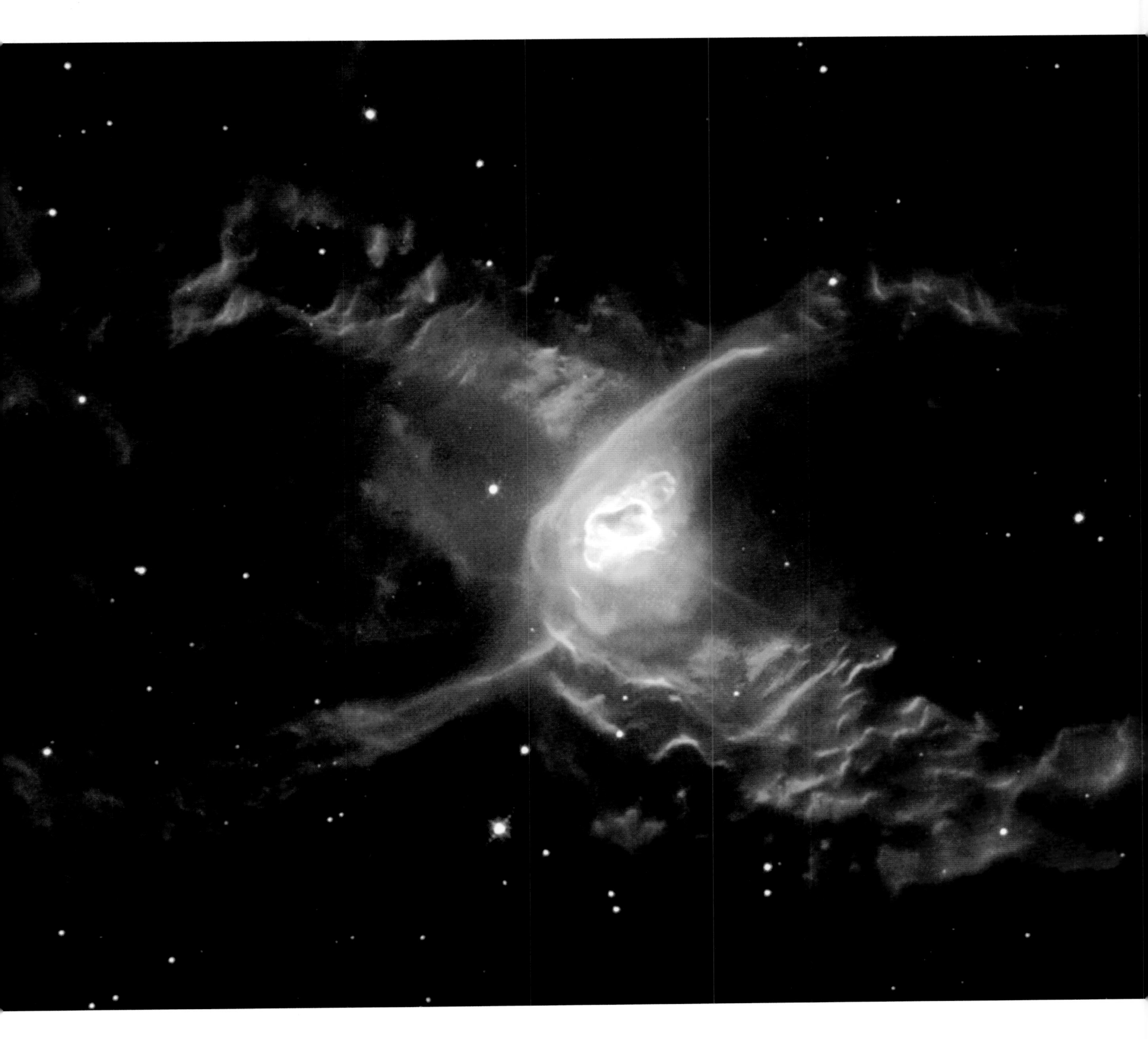

Red Spider Nebula

BRUCE BALICK—WHO CAPTURED THIS IMAGE of the Red Spider Nebula with the Hubble Space Telescope—came up with yet more cosmic poetry when he saw his results, "Oh—what a tangled web a planetary nebula can weave." Some 4,000 light-years from Earth, the Spider is driven by one of the hottest dying stars known. Like many planetaries, it is in a double-star system, and the winds from the stars gust at over 621 miles per second/1,000 km/s. These buffet the surrounding gas and dust, creating the arachnid structures we see here.

Supernova 1994D

A WHITE DWARF STAR EXPOSED at the center of its Planetary Nebula can suffer an even worse fate than losing its atmosphere. In 1994, the Hubble Space Telescope discovered a star exploding in the Galaxy NGC 4526, 55 million light-years away. This supernova—a giant thermonuclear cataclysm—happened to a white dwarf orbiting a companion star. Its nearby partner—attracted by the dwarf's powerful gravity—dumped matter on the star, increasing its mass. Eventually, the dwarf buckled under the strain. It died in a blaze of light (lower left), which rivaled the entire output from the billions of stars in the galaxy.

Supernova 1987A

ONE HUNDRED SIXTY THOUSAND LIGHT-YEARS AWAY, Supernova 1987A in the Large Magellanic Cloud—our major neighboring galaxy—was the nearest to be discovered in recent years and the closest to be observed for the first time with modern instruments. It exploded on February 23, 1987, and was easily visible in the skies of the southern hemisphere for several months. It made a beautiful contrast with the star-forming region of the Tarantula Nebula (left). Unlike 1994D, this star blew up because it was supermassive, unstable, and had expended all of its nuclear fuel.

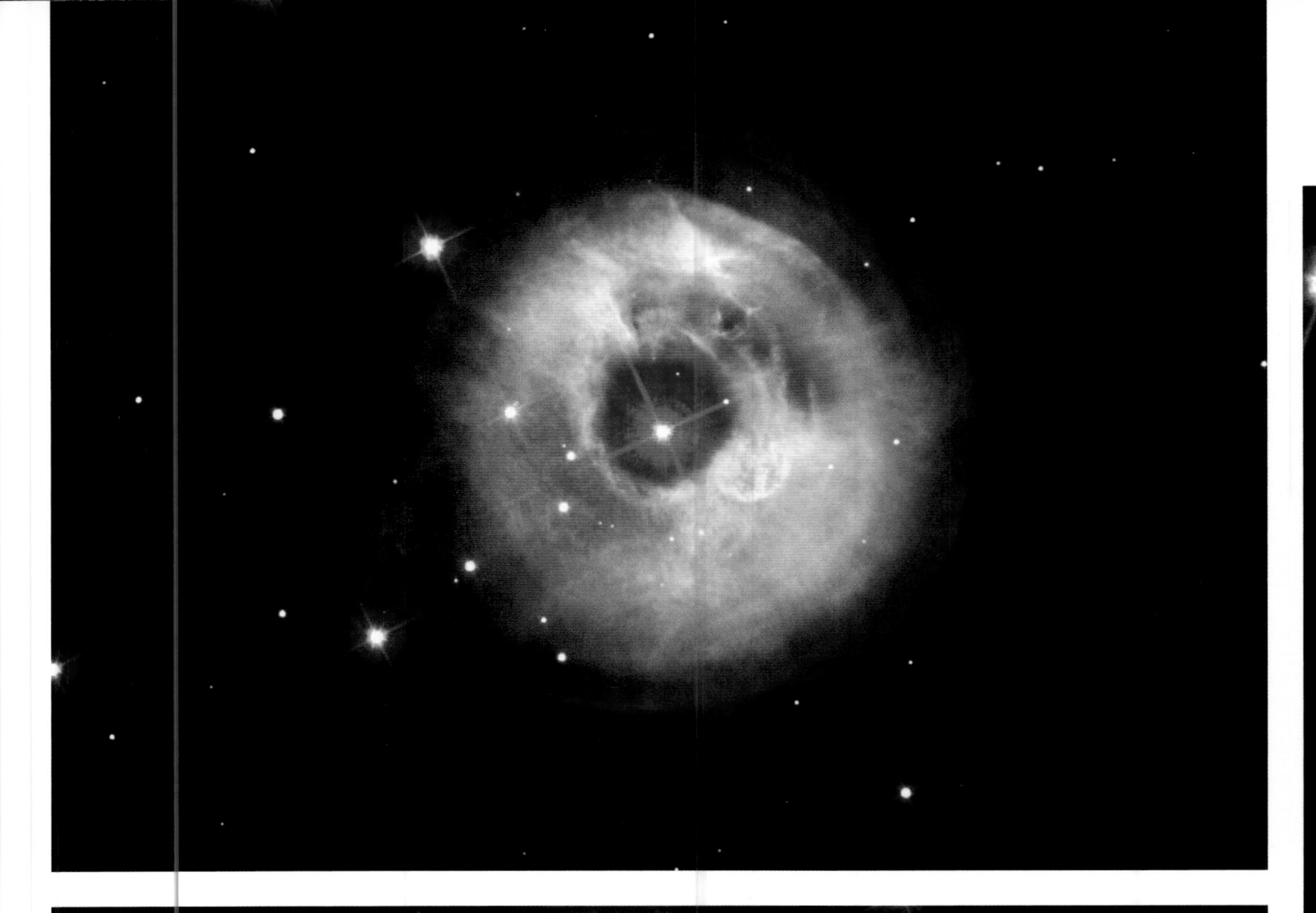

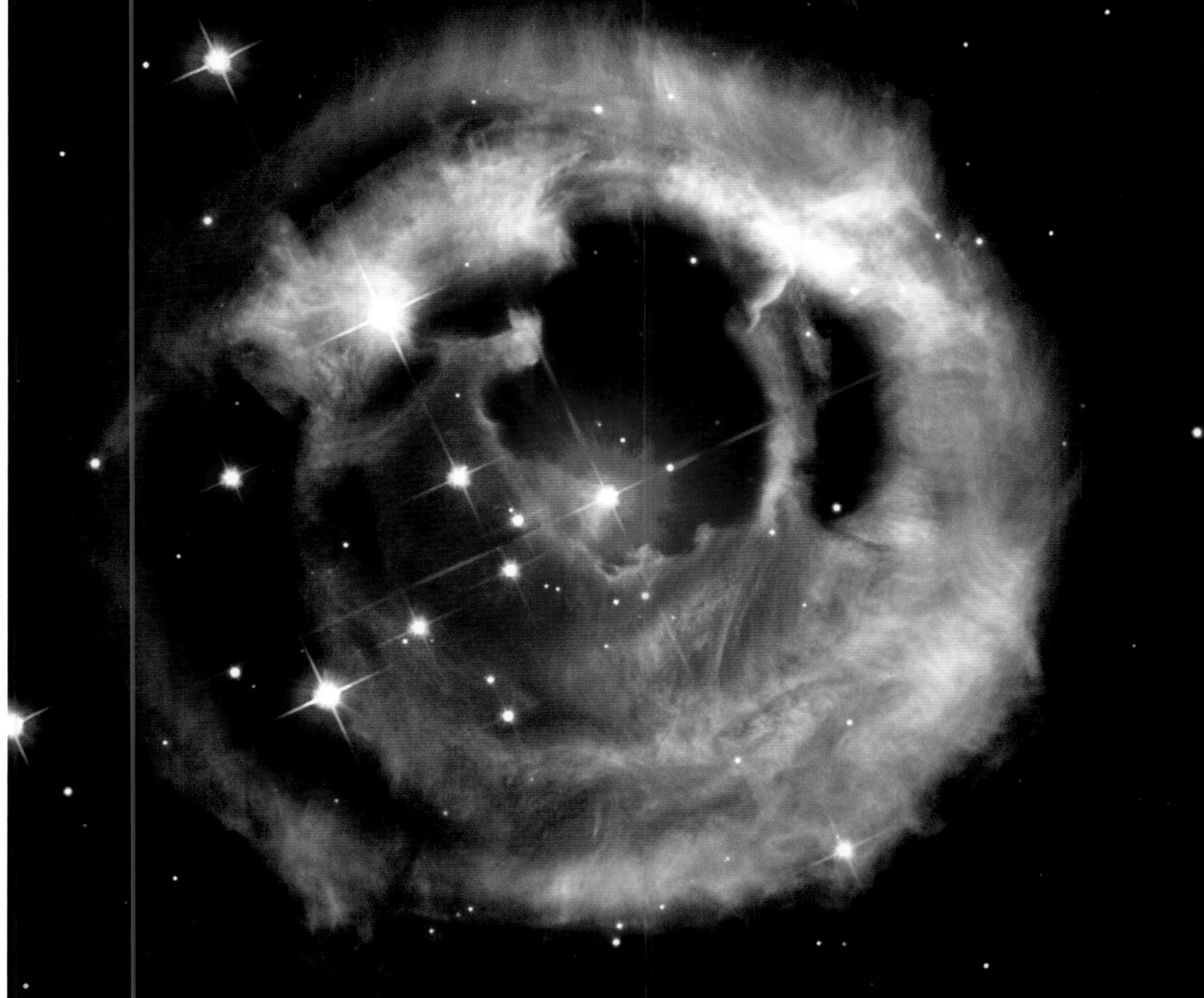

Light echoes

TWENTY THOUSAND LIGHT-YEARS AWAY—at the very edge of our galaxy—the star V838 Monocerotis underwent a sudden outburst in January 2002. It became 600,000 times brighter than the Sun, making it the most luminous star in the Milky Way. These three images from the Hubble Space Telescope chronicle the progress of the flare-up, which researchers still cannot explain. The rings are "light echoes"—light from the outburst reflected off ejected dust. The star has now faded into obscurity, but since it is an aged red giant, it may destroy itself in a supernova explosion in the relatively near future.

Supernova remnant LMC N49

LOOKING LIKE A COSMIC FIREWORK DISPLAY, N49 is the remains of a supernova that exploded in the Large Magellanic Cloud thousands of years ago. At its heart is the corpse of the original star: a supercondensed neutron star spinning once every eight seconds. In March 1999, researchers picked up a huge blast of gamma rays—the most energetic radiation in the universe—from N49. Probably originating in the neutron star, it saturated their detectors. This discovery led to a whole new area of astronomy: the science of gamma ray bursters.

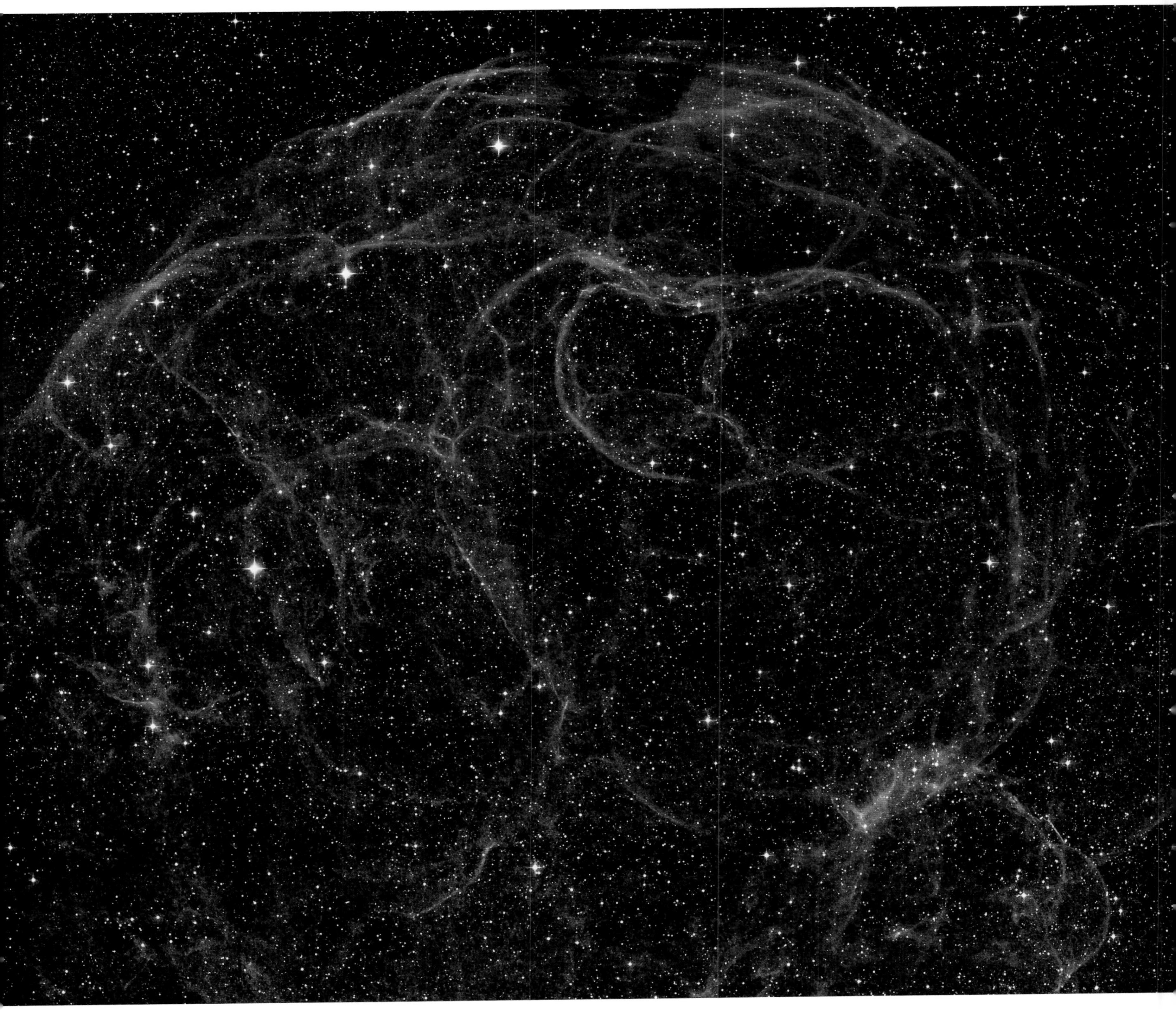

Supernova remnant Simeis 147

THE CONSTELLATION OF TAURUS (THE BULL) is home to this enormous supernova remnant: Simeis 147. Spanning six Moon-widths in the sky, it measures 150 light-years across, and it is still expanding. This image, which reveals delicate filaments and tendrils blasted out from the former star's body, was taken entirely in the light of glowing hydrogen gas. All that remains of the once-brilliant star is its compressed core—a fast-spinning neutron star, or pulsar.

Vela Supernova remnant

ELEVEN THOUSAND YEARS AGO, in the southern constellation of Vela, a mighty star exploded. Being relatively close to Earth—around 800 light-years away—it must have shone 250 times brighter than the planet Venus. It would have been an incredible sight to the first human farmers. Today, the wreckage is strewn over an area sixteen times that of the full moon, and material from the dead star is still expanding into space at thousands of miles per second. The delicate traceries result when the speeding matter hits the cold gas of space.

Pencil Nebula

A CLOSE-UP OF THE VELA SUPERNOVA REMNANT. The long, straight feature—just one light-year in length—is the Pencil Nebula. It was identified by John Herschel (son of William, who discovered Uranus) in the 1840s. The nebula is a shockwave, moving from right to left across this Hubble image. Here, expanding gases from the explosion have encountered a dense region of material in space. The gases, color-coded red (hydrogen) and blue (oxygen), may eventually compress their surroundings to trigger star formation.

The Cygnus Loop

THESE EXQUISITE FILAMENTS FORM PART OF THE CYGNUS LOOP: a complex of cosmic shrapnel ejected from a star that exploded between 5,000 and 8,000 years ago. Even today, its gases are moving outward at a speed of more than 373,000 miles per hour/600,000 kph. As a result, the Loop is vast: even though it is 1,500 light-years away, it still appears more than six times bigger than the full moon in the sky. Coincidentally, William Herschel (whose son John observed Vela), made a study of the Cygnus Loop in the eighteenth century, but it needed the advent of photography to reveal its true nature.

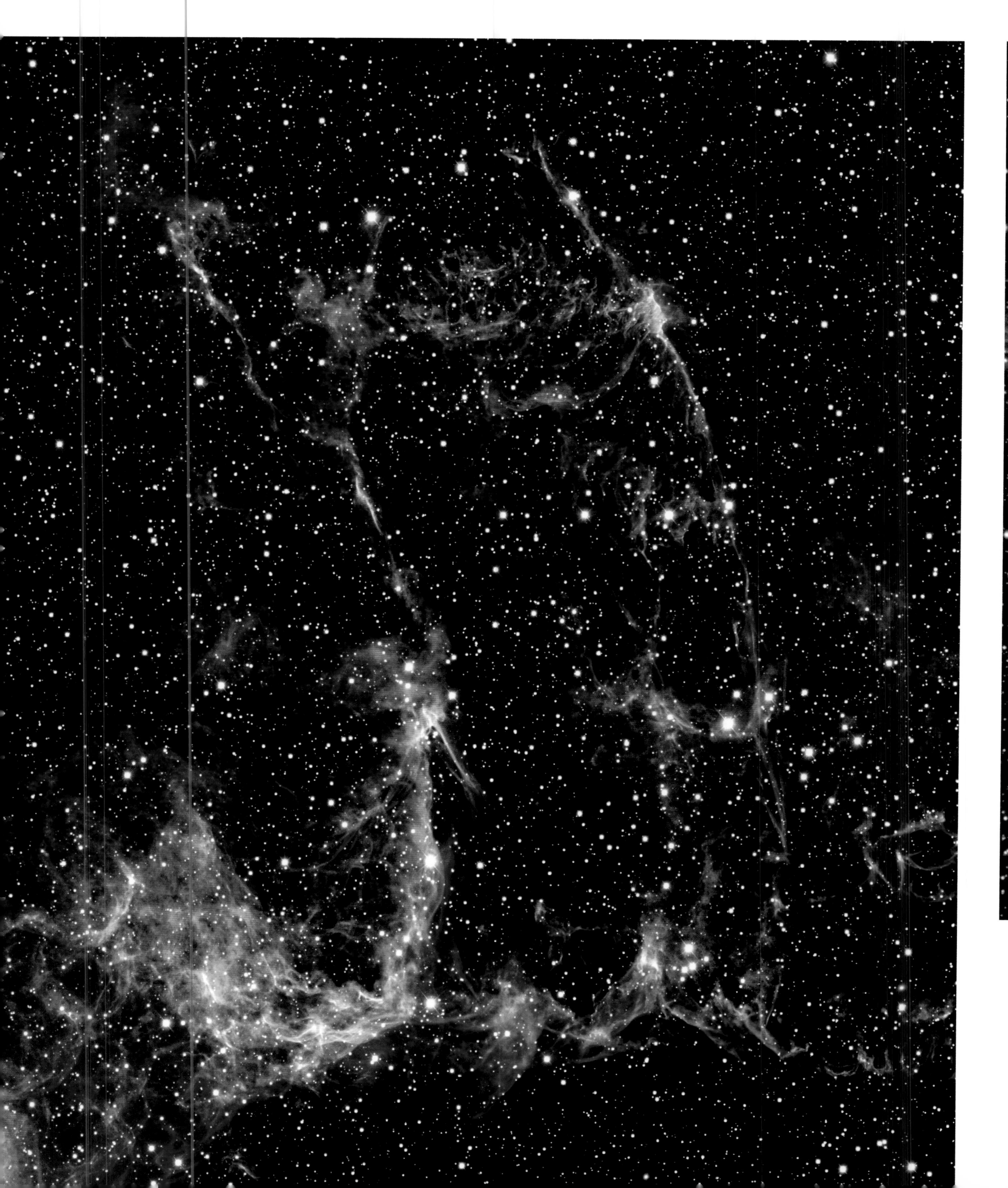

Veil Nebula

PART OF THE GIANT CYGNUS LOOP, the Veil Nebula is one of the loveliest of all the beauty spots in the cosmic landscape. But, like other delicate supernova filaments, it also works as a detective for astronomers. Space between the stars is not empty: it is filled with gas and sooty dust poised to form into future generations of stars. The problem is that the "interstellar medium" is dark and difficult to probe. But when shockwaves from supernovas light up the material—as in the case of the Veil—its structure shines forth, enabling astronomers to understand the nature of star birth.

Crab Nebula

BY FAR ONE OF THE MOST UNUSUAL—and unique—supernova remnants in the sky is the Crab Nebula in Taurus. It is the outcome of a cosmic catastrophe witnessed in 1054 by Chinese, Japanese, and native North Americans. We know from records that the exploding star was visible for a whole year. Now it is a tangle of fiercely heated filaments, expanding at breakneck speed into space. The eerie blue glow at its center is caused by a central pulsar—the core of the old star, compressed to 6 miles/10 km across. It spins at the frenetic rate of thirty times a second, and its powerful magnetic field whips up the surrounding electrons into a radiation frenzy.

NGC 2403/SN 2004dj

IN 2004, THE GALAXY NGC 2403—11 million light-years away—played host to the nearest and brightest supernova visible for over a decade. The exploding star—SN 2004dj—is brilliantly visible at the far top-right (the other bright star is in the foreground). NGC 2403 is a galaxy rich in gas, actively forming young, massive stars. SN 2004dj was a blue supergiant star that ripped through its fuel and exploded after only a few million years—as compared our less massive Sun, which will eke out its reserves and grow old gracefully. It will leave a remnant rather like the Crab Nebula.

W49B

THIS MULTICOLORED MARVEL—courtesy of the 16-foot/5-m telescope on Palomar Mountain and the orbiting Chandra X-ray Observatory—may be a cosmic first. It is definitely a supernova remnant, 35,000 light-years away, but researchers are coming to the conclusion that an unusual gamma-ray burst triggered the explosion. Astronomers have observed these bursters in other galaxies but not yet in our Milky Way. The barrel-shaped relic is incandescent with blue light from energetic X-rays, while atoms of iron and nickel from the catastrophe litter the scene. The former star's core has almost certainly collapsed to become a black hole.

Supernova remnant Cassiopeia A

A FORMIDABLE TRIO OF THE WORLD'S LEADING SPACE OBSERVATORIES—the Hubble Space Telescope, Spitzer, and Chandra—ganged up on the supernova remnant Cassiopeia A to create this composite masterpiece. The coolest, red areas are images from Spitzer, which detects matter around room temperature. Next comes Hubble—it homes in on gas at temperatures of about 18,000 degrees Fahrenheit/10,000 degrees Celsius (coded yellow). Chandra looks at the hottest temperatures of all (shown blue/green), which can reach 18 million degrees Fahrenheit/ 10 million degrees Celsius. But there is still a mystery to be solved. Cassiopeia A's expansion rate tells us that it is nearby, and that its unlucky star must have exploded about three hundred years ago. So why did no one report it?

THE ROMANS CALLED IT THE "VIA LACTEA"—the "Road (or Way) of Milk." In those days before light pollution, thousands of years ago, the Milky Way must have been an awesome sight arching across our skies.

The softly glowing band of stars that makes up the Milky Way is a view of our home galaxy, seen from the inside. The galaxy is shaped like a disk—and we live within that disk. Nearby stars are scattered all over the sky, but perspective makes the more distant stars concentrate into a river of light spanning the heavens.

Our local star city is home to 200 billion stars, and it measures 100,000 light-years across. The Sun is just one of them, living quietly out in the galactic suburbs. There is nothing to single out our Sun from any of the other stars, and we now know that many others have planets where life might exist.

The galaxy has several tiny companions—dwarf galaxies—which will eventually merge with the Milky Way. But it also boasts two major satellites, the Large and Small Magellanic Clouds. These neighbor-galaxies are a stunning sight in the skies of the southern hemisphere, looking like detached portions of the Milky Way itself.

If you were able to swoop over our galaxy and see it from above, you would be treated to the most exhilarating sight—a delicate spiral of stars, like a cosmic Catherine wheel. The spiral arms are blue: they are largely made of hot young stars that blaze at temperatures of up to 54,000 degrees Fahrenheit/30,000 degrees

Celsius. The blue is interspersed with blotches of red: regions where fledgling stars have just been born out of the abundant cosmic gas. Dark sooty dust makes a dramatic contrast with the luminous arms. This is matter-in-waiting: the stuff poised to form into the next generation of stars.

In contrast, the central hub of our galaxy shines yellow—a haunt of aging stars, possessing little in the form of building materials. But researchers now have evidence that our galactic nucleus has fire in its belly. At its heart is a supermassive black hole. This has flared up in the past and created enormous explosions—as well as some weird structures in the downtown regions of our Milky Way.

Enveloping the galaxy is a tenuous, spherical region called the halo. It appears to be largely empty, populated only by a few hundred compact balls of stars called "globular clusters." These globular clusters may provide the key to how our galaxy formed. Astronomers now believe that they are the building blocks of galaxies—and that studying these ancient edifices will give them clues into the history of our Milky Way.

And there is more to the halo than meets the eye. Our galaxy spins, and its rate of spin reveals that it is much more massive than can be accounted for by adding up all the stars we can see. The culprit is "dark matter," which lurks in the halo. At present, we can only guess at the nature of dark matter, but when it is found, its discovery will be a breakthrough in astrophysics.

7 THE MILKY WAY

Previous pages | Northern Milky Way

THIS AWESOME IMAGE of the whole of the northern Milky Way was taken by renowned astrophotographer John Chumack. Observing from a meadow 11,400 feet/3,400 m up in Colorado, near Copper Mountain, Cormack used a Pentax 90mm lens to capture the spectacle. The panorama needed four separate shots, each thirty seconds in duration, which were then digitally stitched together. In this glorious view, we see brilliant young stars, glowing nebulae, and dark dust clouds destined to give birth to the next generation of stars. It is a celebration of a galaxy that is still in its prime.

Milky Way: a side view

IF YOU COULD SEE OUR GALAXY from the side, it would look like two fried eggs lying back-to-back. This artist's impression of our Milky Way seen from its edge reveals its true proportions: a slender disk surrounding a nuclear bulge. The "white" of the egg—our star city's disk, where we live—is incredibly thin. Its central bulge—the "yolk"—is around 6,000 light-years thick. The two regions are very different. The disk is largely made up of young stars, while those in the nucleus are older. Surrounding the galaxy is a spherical halo, containing more than one hundred globular clusters (shown as dots).

Milky Way: a view from the southern hemisphere

SEEN FROM THE CLEAR SKIES of the southern hemisphere, the galactic bulge dominates this photograph. It is centered on the constellation of Sagittarius, whose mighty star clouds mask the location of the galaxy's heart. The nuclear bulge extends for about 10,000 light-years and consists of much older stars than those in the disk. These ancient stars are cool. They shine red or yellow—imparting a warm, yellowish glow to the downtown regions of our Milky Way.

Milky Way in Scorpius

LIVING AS WE DO, in our galaxy's disk, perspective makes more distant stars appear in the sky as a band. In 1610, Galileo—using the newly invented telescope—was the first person to find that our Milky Way was made of "congeries [clumps] of stars." Later, in the eighteenth century, William Herschel surmised that our galaxy had a shape like a grindstone or a lens. This photograph shows the band of the Milky Way passing through the constellation of Scorpius.

Milky Way in Sagittarius

IN THE NEIGHBORING CONSTELLATION OF SAGITTARIUS, billowing star clouds make a dramatic contrast with the blackness of space. Intermingled with the brilliant stars are minuscule grains of dust—"soot" blown off the cooling surfaces of old stars. These show up as dark patches against the glowing backdrop. In some of the denser concentrations, young stars may be forming.

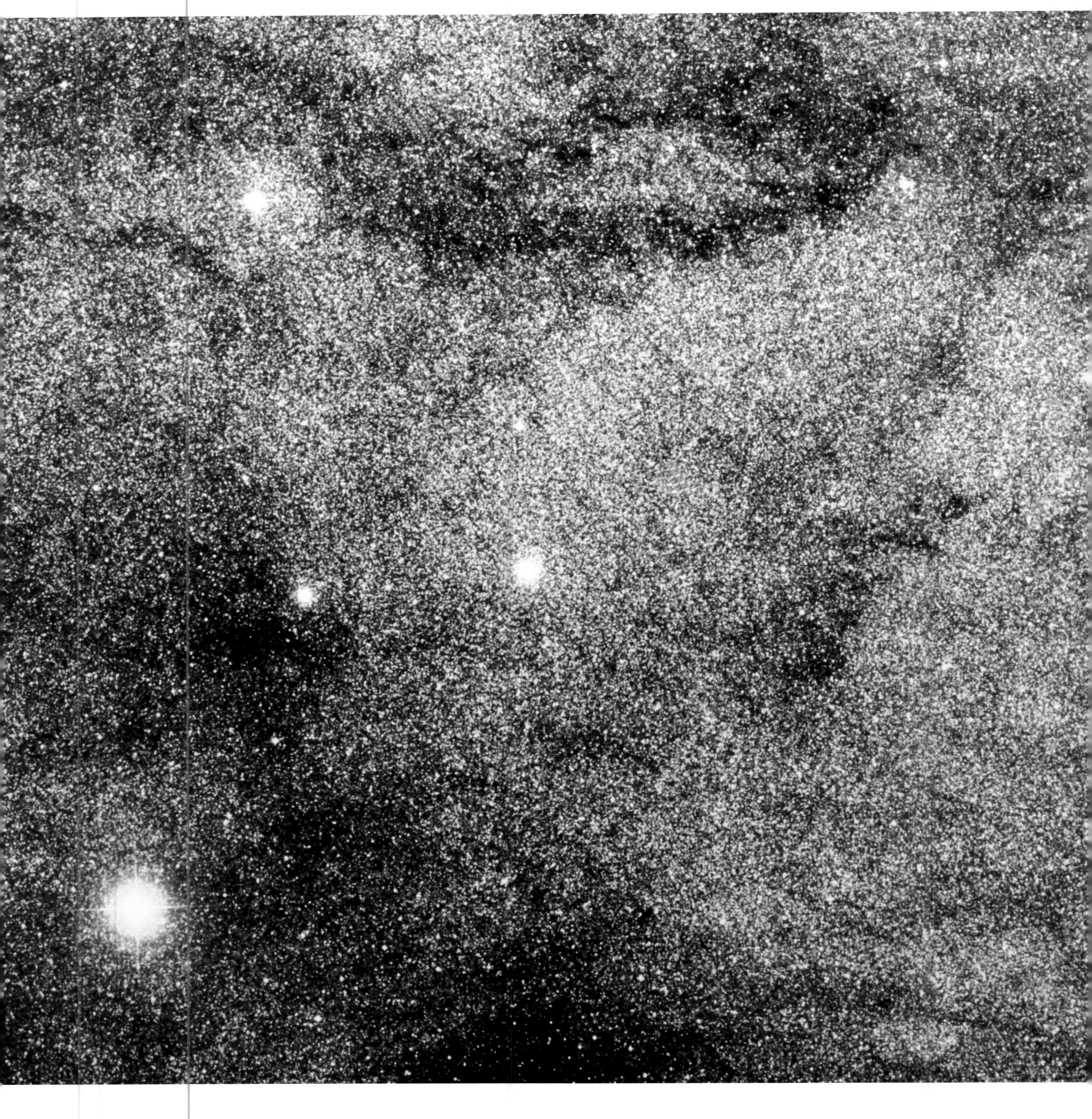

Stellar field

BETWEEN THE CONSTELLATIONS OF SCORPIUS AND SAGITTARIUS—in the direction of the galactic center—lies this extremely crowded star field. The two bright stars in the image are much closer to Earth. Gamma Sagittarii (lower left) is a medium-bright giant yellow star. W Sagittarii, above, is a variable star whose light output changes as it swells and shrinks every 7.6 days. In between—in the middle of the image—are two remote globular clusters. These ancient balls of stars were among the first objects to be born in our galaxy.

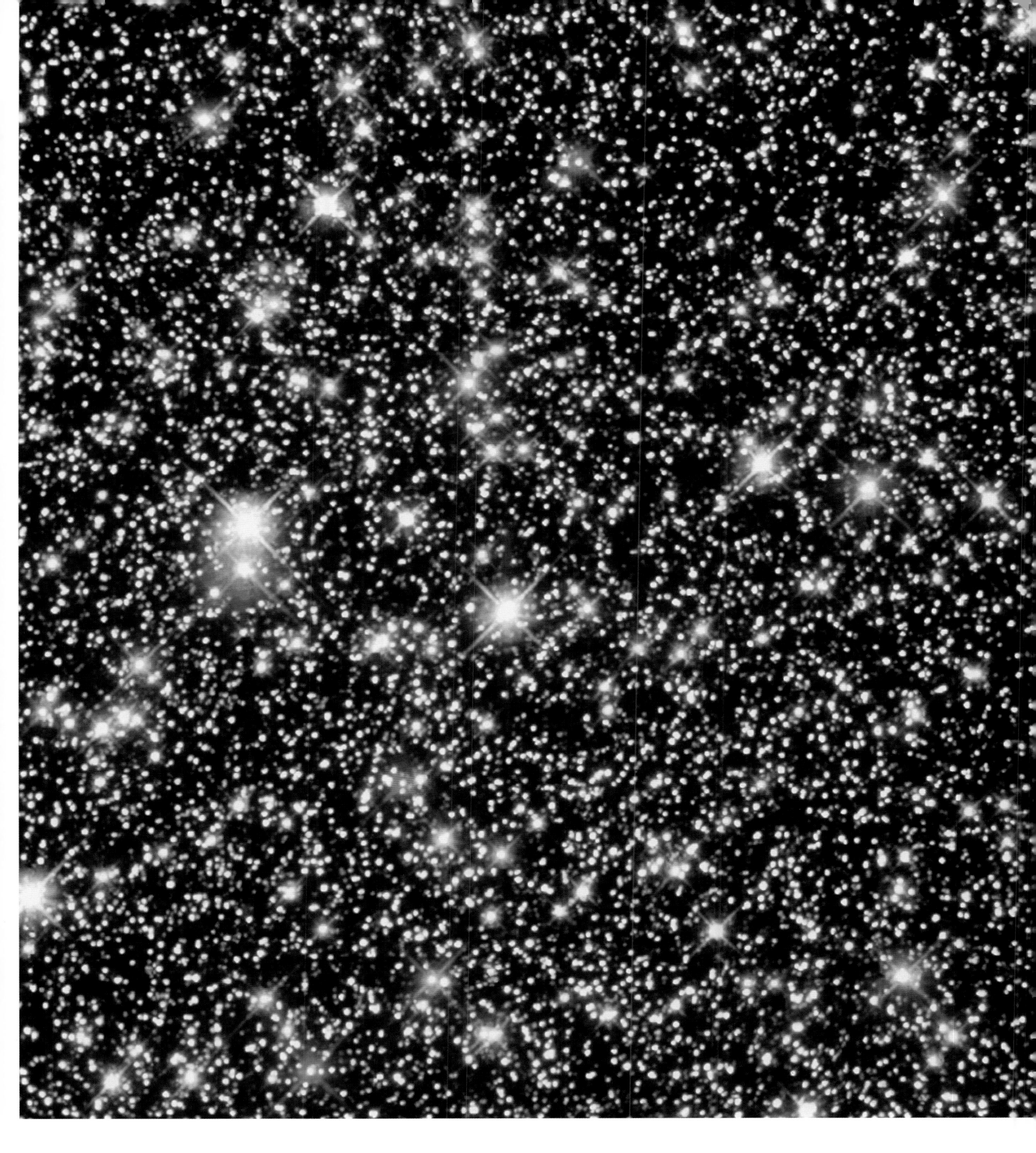

Hubble star image

THE HUBBLE SPACE TELESCOPE captured this image of the myriad stars in the direction of our galaxy's heart. The colors of the stars speak volumes. Red stars are the coolest—their surfaces glow at around 3,600 degrees Fahrenheit/2,000 degrees Celsius. Many of them are old red giants on the way out. Yellow stars, like our Sun, shine at 11,000 degrees Fahrenheit/6,000 degrees Celsius. At the top of the scale are white and blue stars, which can reach temperatures of 54,000 degrees Fahrenheit/30,000 degrees Celsius. Some of these stars are supergiants, whose nuclear reactions are running out of control. They may even explode as supernovas.

Milky Way in the night sky

THIS VIEW OF THE MILKY WAY TOWARD SAGITTARIUS reveals heaped-up clouds of dark, obscuring dust. But this is not the only dark matter in our galaxy. In recent years, researchers have discovered—from the way stars and galaxies move—that there is a great deal of invisible matter in the universe. Astronomers now believe that 90 percent of our own galaxy is made of "dark matter," but they still don't understand what it is made of—or how to detect it.

Milky Way running through the Summer Triangle

UNDER THE DARK SKIES OF ARIZONA, the Milky Way wends its way through the stars of the Summer Triangle down to the horizon. The three stars are Vega in Lyra (bottom right); Deneb in Cygnus (top right); and Altair in Aquila (by the trees, lower left). Only in places like this, with no light pollution, can you see the true glory of the Milky Way arching across the sky. In cities—where glaring lights reflect off particles of dust and debris—viewing the Milky Way is a thing of the past.

Milky Way in Cygnus ▷

ANOTHER VIEW OF OUR LOCAL ARM, as it flows through the constellation of Cygnus, the Swan. The dark cloud is the "Cygnus Rift"—which is very obvious to the unaided eye. Brilliant Deneb (top) is brighter than 100,000 Suns. To its left is the aptly named North America Nebula, a region of star formation. The stars that mark the "arms" of the "Northern Cross" are at the top right and lower left. Below the star on the left is the Cygnus Loop, a huge circular supernova remnant. It is a grim reminder that star death accompanies star birth—especially when it involves massive, unstable young stars.

Star clouds in the Milky Way

THESE GIANT STAR CLOUDS on Aquila lie about five hundred light-years away and are part of the spiral arm of the galaxy that we live in. On this side of our Milky Way, there are three major spiral arms. Innermost is the Sagittarius Arm, then there is our Local Arm, and outside is the Perseus Arm. Our Local Arm also contains the huge Orion star factory, the beautiful star clusters of the Pleiades and the Hyades, and Sirius—the brightest star in the sky. Like all spiral arms, it is a hotbed of star formation.

Milky Way from the southern hemisphere

FROM THE SOUTHERN HEMISPHERE, the Milky Way is a dazzling sight. Here, you are looking in the direction of the galactic center, where the stars appear most densely packed. The brilliant red gas cloud in this image is the Carina Nebula, which surrounds the erratically varying star Eta Carinae. This star—estimated to be over one hundred times heavier than our Sun—will probably explode as a supernova in a few thousand years' time. On the lower right is our biggest satellite galaxy, the Large Magellanic Cloud.

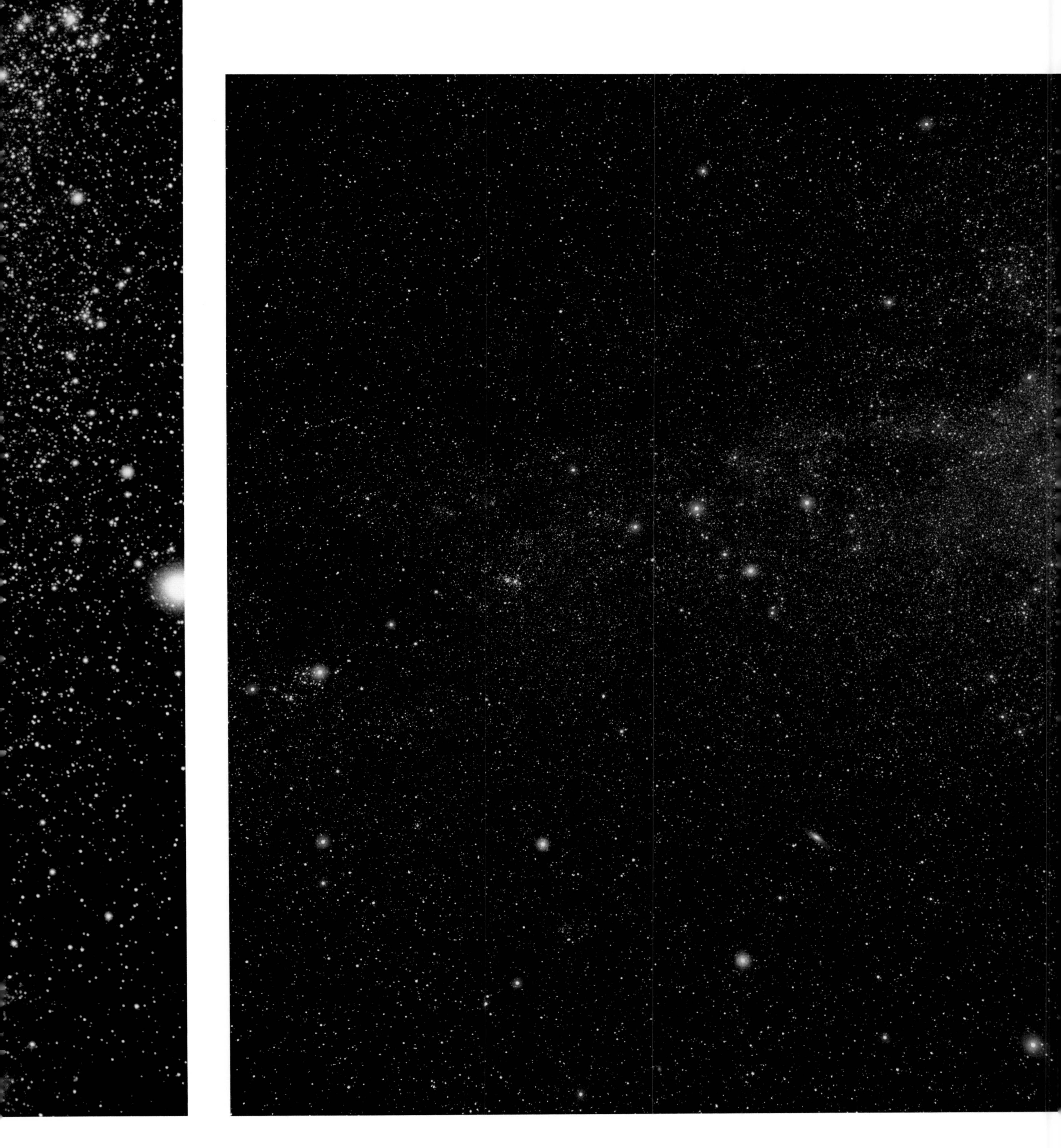

Northern Milky Way

ALTHOUGH THE NORTHERN MILKY WAY looks fainter than its southern regions—because you're looking outward, rather than inward—it nevertheless contains some fascinating objects. The line of stars on the left is the constellation of Perseus, with the "Double Cluster" of young stars at its right-hand end (center of image). The stars in these clusters are just 3 and 5 million years old, respectively—as compared to our middle-aged star's 4.6 billion years. Next comes the W-shaped constellation of Cassiopeia. The small streak on the lower right is the Andromeda Galaxy. At 2.9 million light-years away, it lies way outside the bounds of our Milky Way.

◁ Milky Way in Scutum

BRILLIANT STAR CLOUDS CONTRAST with pitch-black lanes of interstellar dust—"soot" from cool stars—in the constellation of Scutum. William Herschel, who discovered Uranus, was the first to make an in-depth study of the Milky Way. One night in 1784, peering through his 20-foot- / 6-m-long telescope, he yelled to his observing companion and sister, Caroline: "Here is certainly a hole in the heavens." Herschel believed that what we now know to be dark material was actually tunnels in space. But Caroline suspected there was a different explanation . . .

Milky Way and Coalsack Nebula

CAROLINE HERSCHEL LOGGED ALL OF HER BROTHER'S "TUNNELS," and when her nephew, John, went off to observe south of the equator, she urged him to find more. The most striking of all is the Coalsack, easily visible in southern hemisphere skies next to the Southern Cross (to the right and slightly above the Coalsack). Although in the West we interpret the region as a "coalsack," the Aborigines in Australia saw it as an emu lying in wait for a possum perched in a tree. This giant cloud of gas and dust contains enough material to make 40,000 stars.

Milky Way

LOOKING TOWARD THE GALACTIC BULGE IN SAGITTARIUS, this image reveals star clouds, nebulae, and dark dust piled upon one another in the line of sight. It also catches a nearby cosmic interloper: Halley's Comet (bottom left). The most famous of all "dirty snowballs," it only sweeps through the inner solar system every seventy-six years. On the last occasion—in 1986—it was a difficult object to see. In the northern hemisphere, it was far away and faint. When it reached the southern hemisphere, it hung, camouflaged, against the Milky Way for days. Only when it emerged from its starry backdrop did it become a spectacular object in southern skies.

Milky Way in Cygnus

THIS IMAGE OF THE MILKY WAY IN CYGNUS covers a very similar area to that on page 197. However, it has been taken through a filter, which allows light from gas—rather than stars—to dominate. The dark Cygnus Rift and the Cygnus Loop (below left) are very obvious. But hydrogen and helium gas at a temperature of 14,000 degrees Fahrenheit/ 8,000 degrees Celsius take center stage in this image. The gas between the stars—unlike the gas in the Earth's atmosphere—is far from uniform. It has four components. Most of it is like the "warm" gas here; interspersed are bubbles of tenuous gas at millions of degrees. Then there are cool filaments of hydrogen, like cirrus clouds. Coldest of all are the dark clouds like the Coalsack—the birthplaces of the next generation of stars.

Milky Way in Cassiopeia and Perseus

ANOTHER VIEW OF THE MILKY WAY, toward its edge, homing in on the constellations of Perseus (left) and Cassiopeia. On the lower left are the Heart and Soul nebulae; the Andromeda galaxy is at lower right. In all cultures, the constellations have been named after ancient legends: in the myths of Ancient Greece, Perseus was a winged superhero who rescued the Princess Andromeda; Cassiopeia was her mother. The W-shaped constellation is supposed to look like the queen sitting in a chair.

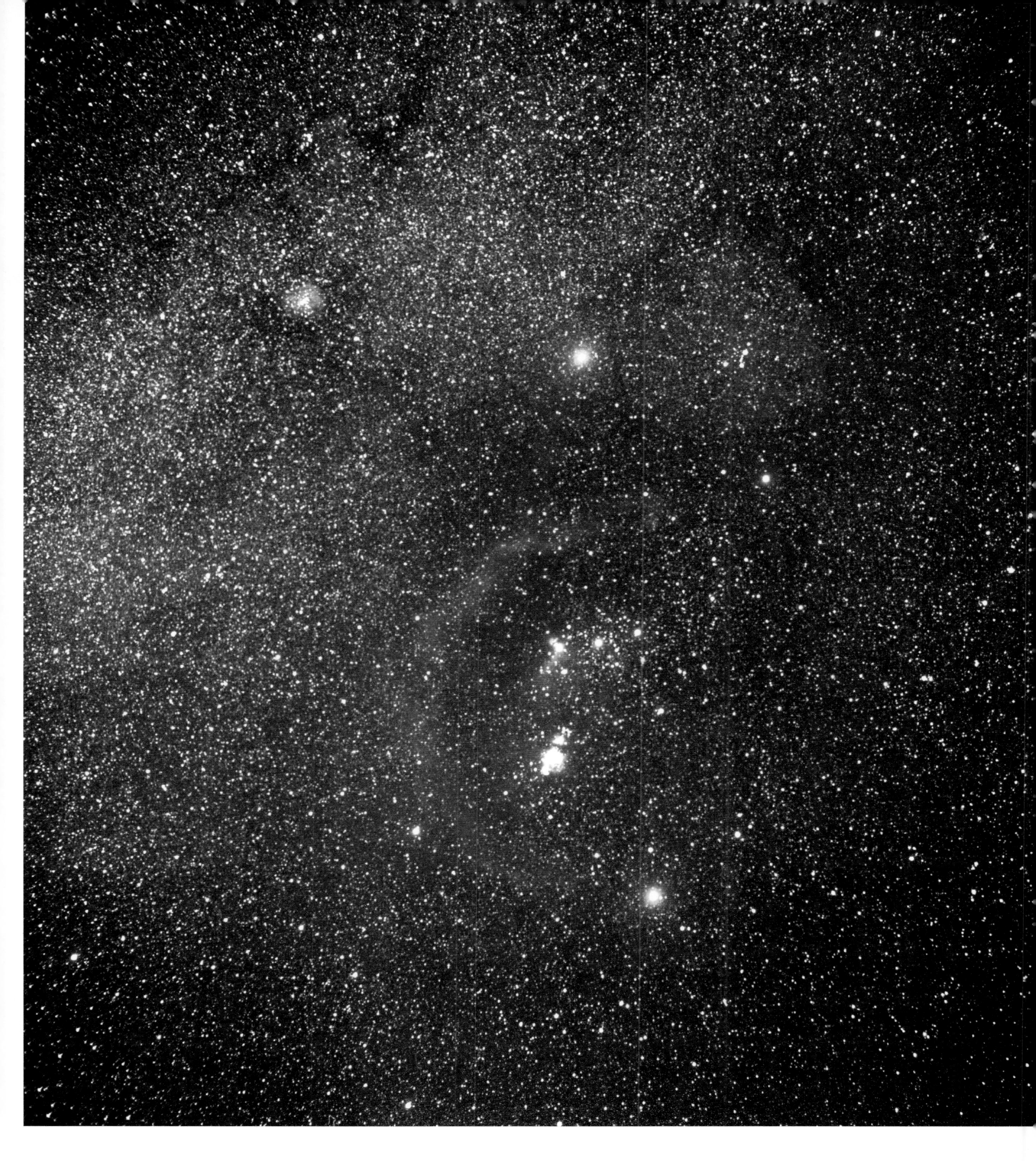

Milky Way from Monoceros to Orion

THE CONSTELLATION OF ORION, fabled as a legendary giant, is captured here in the red light it emits. On the top left is the Rosette Nebula, while the Orion Nebula—another active region of star formation—appears just below the center, beneath the three stars of Orion's belt. The huge arc that encircles the constellation is Barnard's Loop. Hundreds of light-years across, Barnard's Loop has been expanding for millions of years. Astronomers believe that giant young stars exploding as supernovas in the Orion region have powered this enormous cosmic soap bubble.

Milky Way in Scorpius and Ophiucus

SEEN FROM COLORADO, the star clouds of the central Milky Way billow down over a dramatic mountain backdrop. Dark drifts of dust and gas—poised to create the next generation of stars—blot out the center of this luminous landscape. The brilliant red star on the right is Antares. Its name means "rival of Mars." This red giant star—over fifteen times more massive than the Sun—is reaching the end of its life. It has been revered by many civilizations in the past, including the Persians and the Egyptians. The astronauts on the Apollo 14 mission in 1971 named their landing module after the star.

Central Milky Way

AS WE CLOSE IN ON THE CENTER OF OUR GALAXY, a frenzy of star formation starts to build. Pink nebulae, burgeoning with newly born stars, dominate the scene. The dark dust and gas clouds—seen here in this image—are biding their time but will later collapse to form new suns. At top right, and much closer, is the rho Ophiuchi complex. Five hundred light-years away, these pink and blue clouds are the nearest star nursery to Earth. The beautiful blue nebulae are the result of interstellar dust reflecting light from young stars—which have only just been born.

Optical image of the Milky Way in Sagittarius

THE CONSTELLATION OF SAGITTARIUS—which marks the location of the core of our galaxy—is dotted with beautiful glowing nebulae: the birthplaces of newborn stars. At the center is the Lagoon Nebula M8. Some 5,200 light-years away, it is a huge star-forming region, named after the dark, lagoonlike rift that runs through its center. Its name—M8—commemorates the eighteenth-century French astronomer Charles Messier. Irritated by fuzzy objects resembling comets (which he wanted to discover), he compiled a catalog of 104 nebulae and galaxies. Ironically, this catalog emerged to become his legacy.

The center of the Milky Way, with Jupiter and Antares

A VIEW OF THE CENTRAL REGIONS OF OUR MILKY WAY, with red Antares (lower right) standing out against the background of distant stars. The interloper at lower left is the planet Jupiter—one of our nearest worlds. Astronomers are now discovering that many of the billions of stars in our galaxy are circled by planets. Technology, so far, has only allowed them to detect massive worlds, like Jupiter, but space missions in the next few years are being designed to seek out planets the size of the Earth—which may be blessed with life.

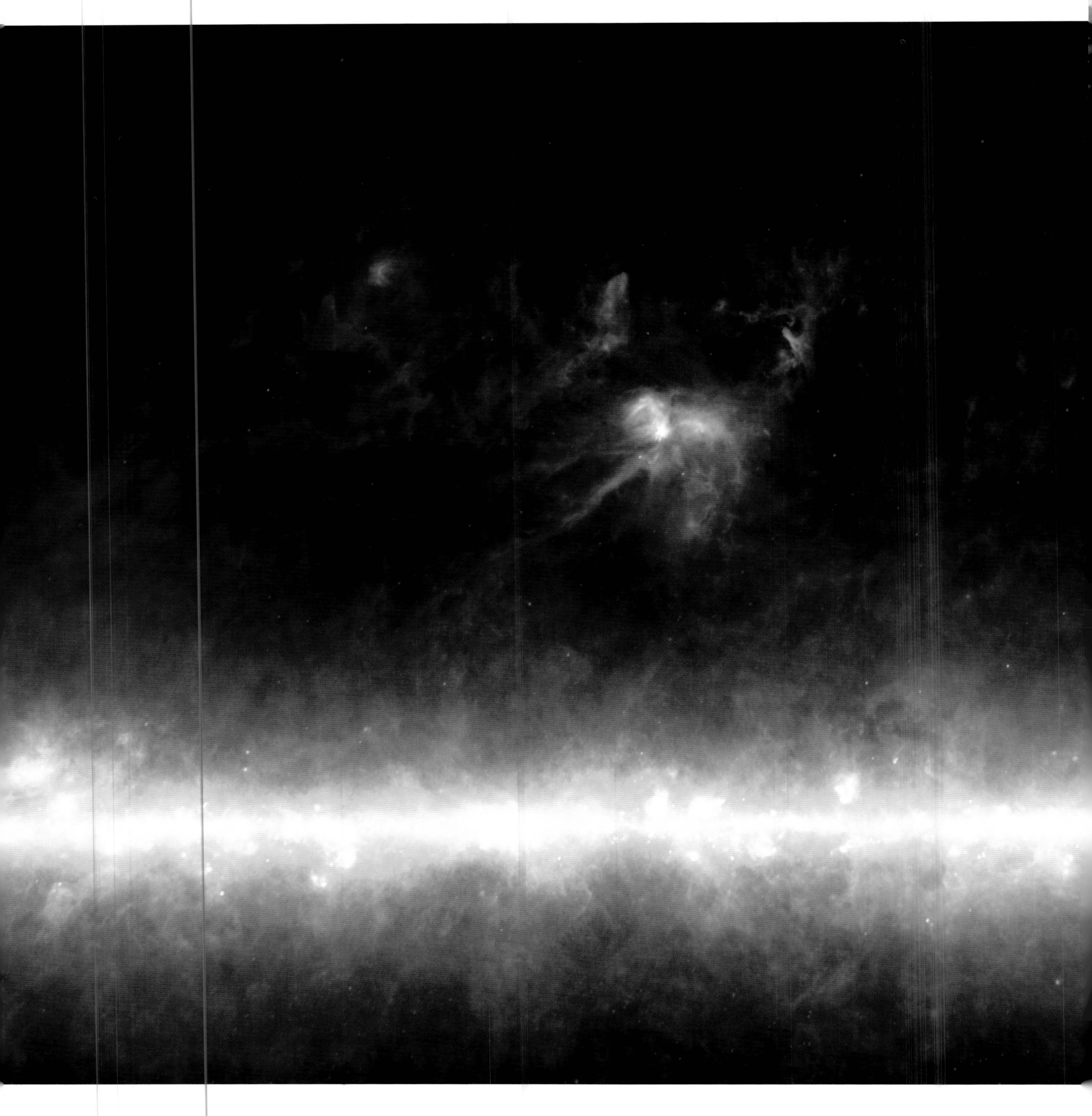

The heart of the Milky Way: infrared image

A TOTAL OF 25,000 LIGHT-YEARS AWAY, the center of our galaxy is an enigma. Astronomers have suspected for many years that it is a violent place, but sooty dust has always obscured the view. Then, in 1983, the Infrared Astronomical Satellite (IRAS) sliced through the dust by latching onto heat radiation instead of light and provided a window into the very heart of our Milky Way. This image reveals hot, disturbed gas at the galactic center (the band bottom of image). Above is the glorious star-forming region of rho Ophiuchi.

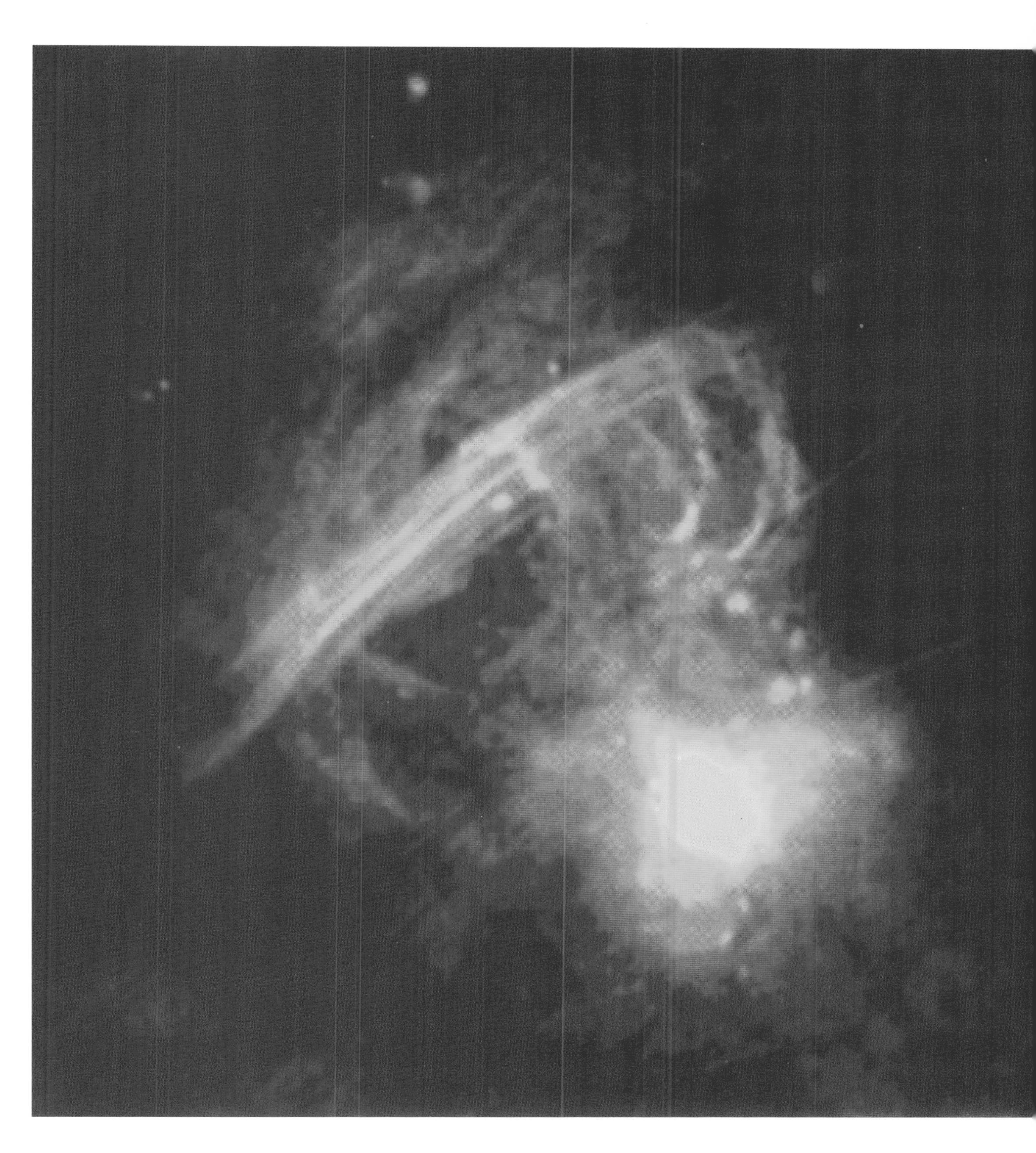

Radio view of Arc and Sagittarius A

DOWNTOWN IN THE MILKY WAY, radio telescopes, working at long wavelengths, don't even notice the small dust particles that obscure the center. This image, captured by the ganged-up forces of the Very Large Array in New Mexico, reveals our galaxy's core. At the center is its powerhouse: Sagittarius A. The brilliant gas almost certainly cloaks a massive black hole (see Chapter 9). Surrounding it are the delicate filaments of the Arc—a huge magnetic structure with tendrils, 150 light-years long but only half a light-year wide.

Galactic center: X-ray image

X-RAY VISION FROM THE CHANDRA SATELLITE provided this image of our galaxy's heart. The region around the black hole (center) is color-coded white. The energy from the black hole—weighing in at nearly 3 million times the mass of the Sun—makes the gas in its neighborhood glow. In this image, red is low energy, green is medium, and blue is high. Like many other galaxies, the core of our Milky Way is a maelstrom of activity. But at a distance of 25,000 light-years, we are safe from the violence.

Large and Small Magellanic Clouds

JUST AS THE EARTH'S GRAVITY HOLDS THE MOON in orbit, so—on a vastly greater scale—the Milky Way holds two large satellite galaxies in thrall. The Large Magellanic Cloud (left) and the Small Magellanic Cloud (right) are a sensational sight in the skies of the southern hemisphere. Both are "irregular" galaxies: they are so much smaller than the Milky Way that they do not have the mass to "grow" spiral arms. The navigator and sailor Ferdinand Magellan was the first person to record the clouds on his around-the-world voyage between 1519 and 1521.

Large Magellanic Cloud

AT 160,000 LIGHT-YEARS AWAY, the Large Magellanic Cloud is the closest major galaxy to the Milky Way. It is teaming with the raw materials of star birth. This image reveals hotbeds of star formation —in particular, the Tarantula Nebula (left). Young stars are often violent and unpredictable—and on February 23, 1987, one of them exploded as a supernova. Even at the distance of the Large Magellanic Cloud, it was visible for ten months. At maximum luminosity, it was 250 million times brighter than the Sun.

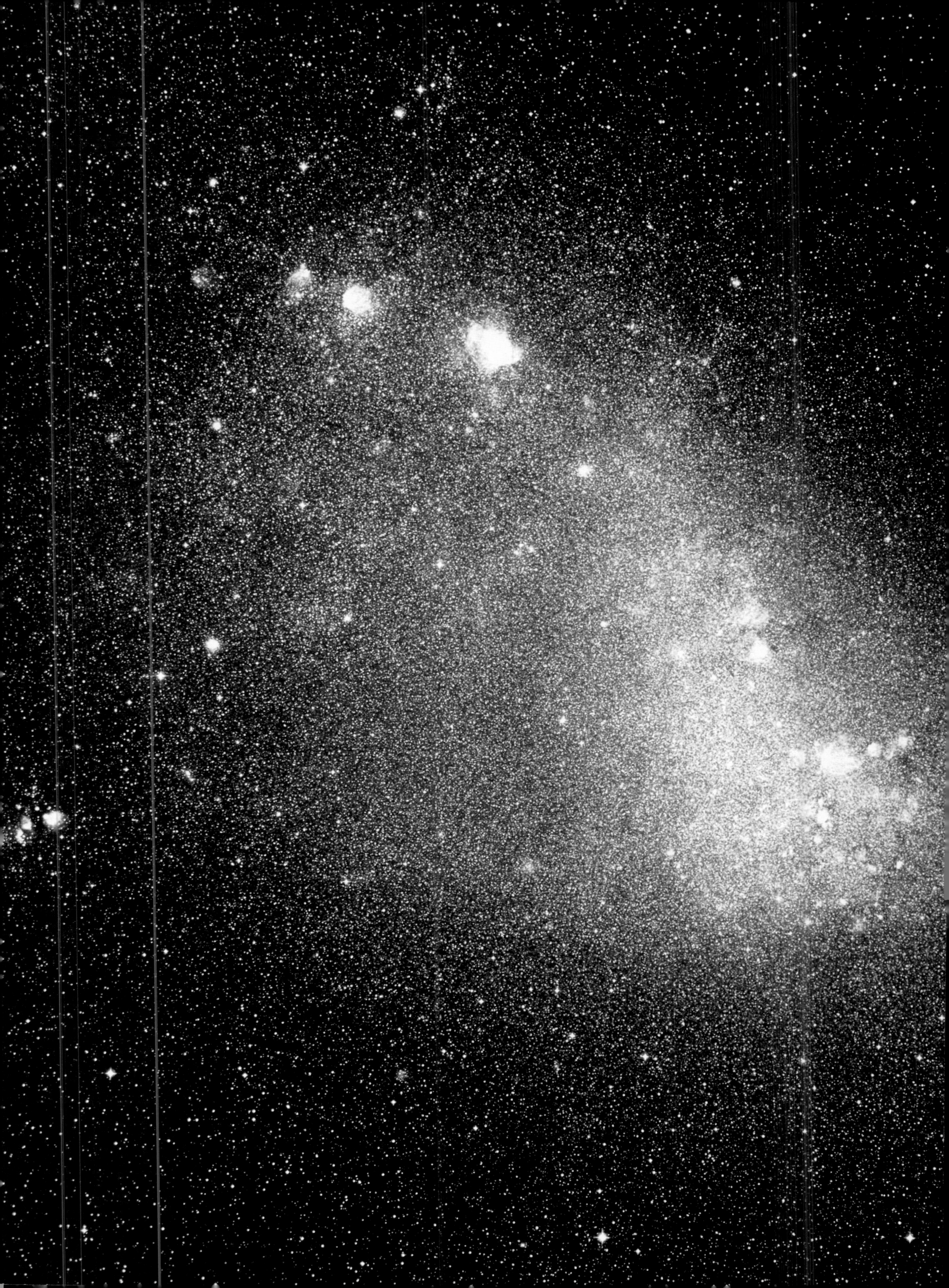

◁ Small Magellanic Cloud

THE SMALL MAGELLANIC CLOUD weighs in at only a quarter of the mass of its companion. Two hundred thousand light-years distant, it is a young, active galaxy busy with star birth. It contains around 2,000 star clusters, many created in a sudden burst of star formation 100 million years ago. In the early twentieth century, the renowned astronomer Henrietta Leavitt observed variable stars in the Small Magellanic Cloud, and used them to establish the distance scale of the universe.

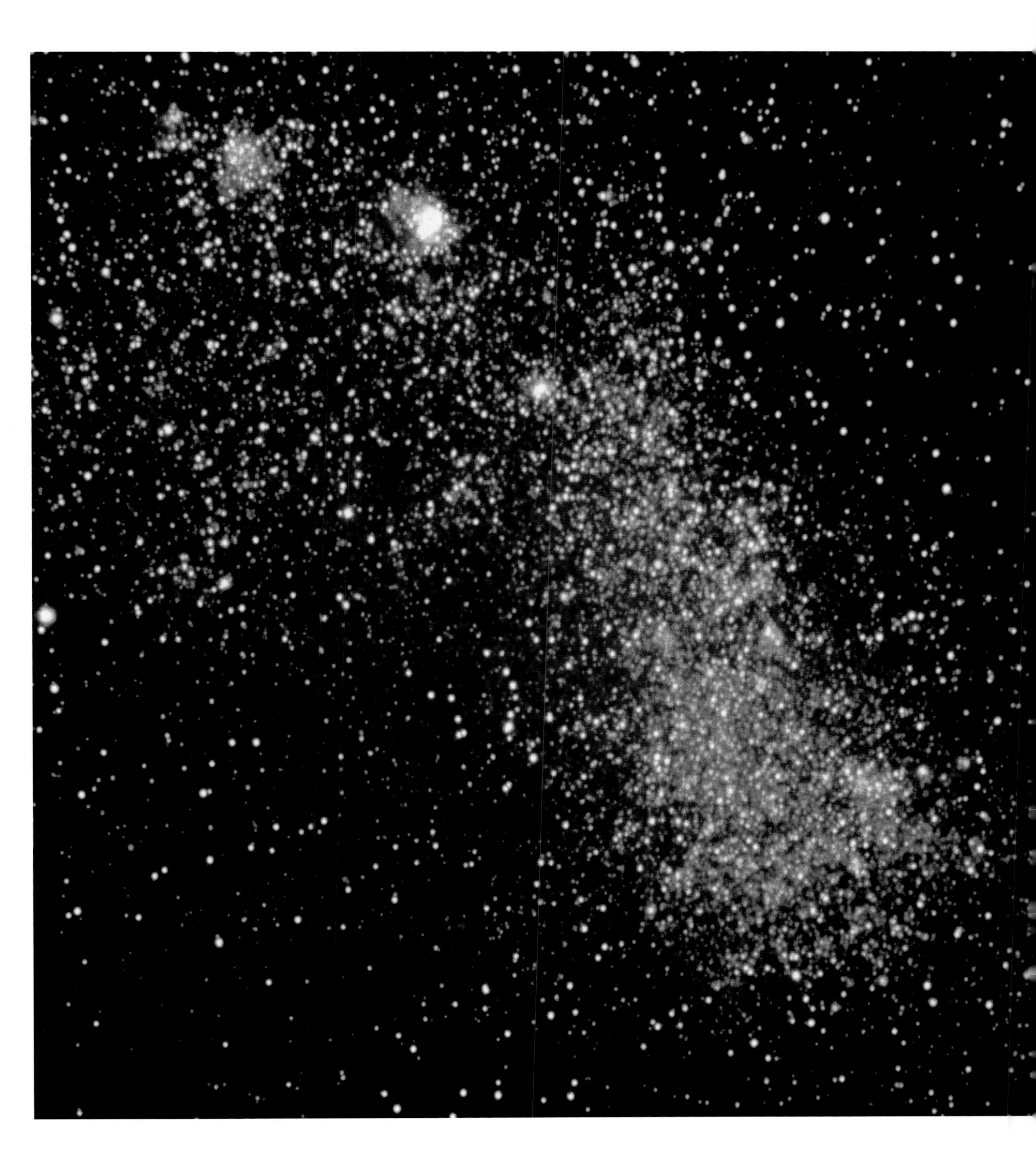

Small Magellanic Cloud

EVERY 1.5 BILLION YEARS, the Large and Small Magellanic Clouds complete their circuit around our Milky Way. But the Small Magellanic cloud suffers, at an enormous price: this is a galaxy being torn apart. The mighty gravity of our Milky Way is taking its toll on the Small Cloud and breaking it up. Its ripped-up debris is strewn around our galaxy as the band of the "Magellanic Stream"—a trail of gas that envelops and links both the Small and Large Clouds. Eventually, this fragile galaxy will merge with the Milky Way, a fate that will befall its larger companion in the future.

"SPIRAL CONVOLUTIONS . . . with successive increase of optical power, the structure has become more complicated"—so wrote astronomer William Parsons, third Earl of Rosse, about the galaxy M51 in 1845, after he had observed the object with his huge telescope at Birr, in the heart of Ireland.

Rosse was the first person to pick out the structure in what were then called "Spiral Nebulas." But at that stage, nobody knew the nature of the beast that he was observing. Were these merely star-forming regions with a complex architecture? Or were they something completely different? Then there were other baffling fuzzy objects, like the Large and Small Magellanic Clouds, which did not fit in with the cosmic scheme of things.

The answer came with the rise of more precise technology and telescopes—and, in particular, with the input of a young American astronomer, Henrietta Leavitt. In the early twentieth century, she detected stars in the Small Magellanic Cloud that varied in brightness like stars in the Milky Way. Knowing the properties of those stars enabled Leavitt to establish a distance to the Cloud. It wasn't a nebula—it was a system of stars way beyond our galaxy.

Enter Edwin Hubble. A trained lawyer, he was a Rhodes Scholar at Oxford and a polymath. He knew he could turn his hand to everything, and he dabbled in astronomy courses in college. Eventually, he abandoned his dazzling legal career for his love of the cosmos. "I would rather be a second-rate astronomer than a first-rate lawyer," he once said.

But second-rate was not Hubble's way. In the 1920s, he followed up on Leavitt's pioneering observations and managed to open up a totally new universe—one that had been hitherto unsuspected. He used the new telescopes on Mount Wilson in California to detect more stars in Rosse's "spiral convolutions." Then he made the most astonishing breakthrough: the discovery that our cosmos is thriving with billions of galaxies like the Milky Way.

Hubble—after whom the famous space telescope was later named—spent the 1930s classifying these new cities in the sky. Many of them are spiral in form, like our Milky Way or its giant neighbor the Andromeda Galaxy. Some are ragged irregulars—visit the southern hemisphere to see the two companion galaxies to our own, the Magellanic Clouds. Then there are bland elliptical galaxies made of old red stars: they are not at all beautiful, but they may be home to a supermassive black hole at the core.

Leavitt and Hubble pioneered our knowledge of the wider universe. But now, with telescopes in space, we understand so much more. Galaxies are gregarious beasts; they cluster together into groups. They are friendly with one another; gravity makes them interact, often leading to a mutual outburst of star formation, which regenerates both galaxies.

Most of all, galaxies help us to excavate the past of our universe. Because they are so far away—which means their light takes so long to get to us—when we look at a galaxy, we are witnessing cosmic archaeology.

8 GALAXIES

Previous pages | Spiral Galaxy M51

SPIRAL GALAXY M51 IS A GLORIOUS CARTWHEEL of stars that encapsulates their birth, life, and death. This classic cosmic city, containing billions of stars, is young and energetic. The pink nebulae are the neon lights of active star formation; the dark dust lanes are pregnant with material to create the next generation. This—the so-called Whirlpool Galaxy—is a beautiful affirmation that our universe has a long way to go.

Distant galaxies

NEEDING 342 CAMERA EXPOSURES, this is one of the deepest images of the universe ever captured. Baby galaxies—all less than a billion years old—tumble over each other in this photograph taken by the Hubble Space Telescope in the mid-1990s. The area targeted by the Hubble was tiny: the equivalent of a tennis ball seen at 328 feet/100 m. Yet it revealed 3,000 objects—most of them galaxies—4 billion times fainter than anything visible to the unaided eye. It set new records in penetrating the cosmos, revealing fledgling star systems quite unlike their grown-up counterparts today.

Spiral Galaxy ESO 269-57

A CLASSIC GALAXY, ESO 269-57 lies 155 light-years away, in a cluster of galaxies in the constellation of Centaurus. Measuring 200,000 light-years across—twice the size of the Milky Way—this galaxy is a spiral like our own. Its inner arms are more tightly wound than those in our galaxy, but its outer regions are very similar to those of the Milky Way. The luminous clumps are shining gas clouds—regions where star birth is actively underway. This exquisite image was taken by the European Southern Observatory's Very Large Telescope in Paranal, Chile.

Spiral Galaxy NGC 2997

NGC 2997 LIES IN ONE OF THE MOST UNLIKELY CONSTELLATIONS—a southern group of stars known as Antlia (the "airpump"—named to commemorate its inventor, the renowned British scientist Robert Boyle). This image from Chile homes in on the galaxy's central bulge. It is home to millions of elderly red and yellow stars, although dust lanes flanking the nucleus indicate that star birth is still on the agenda. The galaxy is about 55 million light-years away.

Overlapping galaxies NGC 3314a and 3314b

THIS SENSATIONAL IMAGE made astronomers believe that they had discovered another Starburst Galaxy, with an eruption of star formation in its core. But astronomer Bill Keel from Alabama realized the true scenario. "It looked like matter blowing off from M82 [see page 257]—but I found from photos taken in Chile that there are two galaxies. They are both in the same galaxy cluster, lined up on one another—but they are too far away from one another to be interacting." The spiral galaxy NGC 3314a is 115 million light-years away; its cohort, NGC 3314b, is 140 million light-years distant.

Spiral Galaxy M83

THIS STUNNING GALAXY, M83, is one of the closest and brightest in the sky. It lies about 15 million light-years away in the constellation of Hydra and is part of a group of galaxies that includes the explosive Centaurus A (see page 256). Like our Milky Way, it is a "barred spiral," having a rectangular wedge of stars across its central nucleus. Rich in gas, it is furiously engaged in star formation—as is evidenced by the hot red gas clouds in its spiral arms. But the downside is that many of the more massive stars die young. So far, astronomers have detected six newborn stars exploding as supernovas in this galaxy.

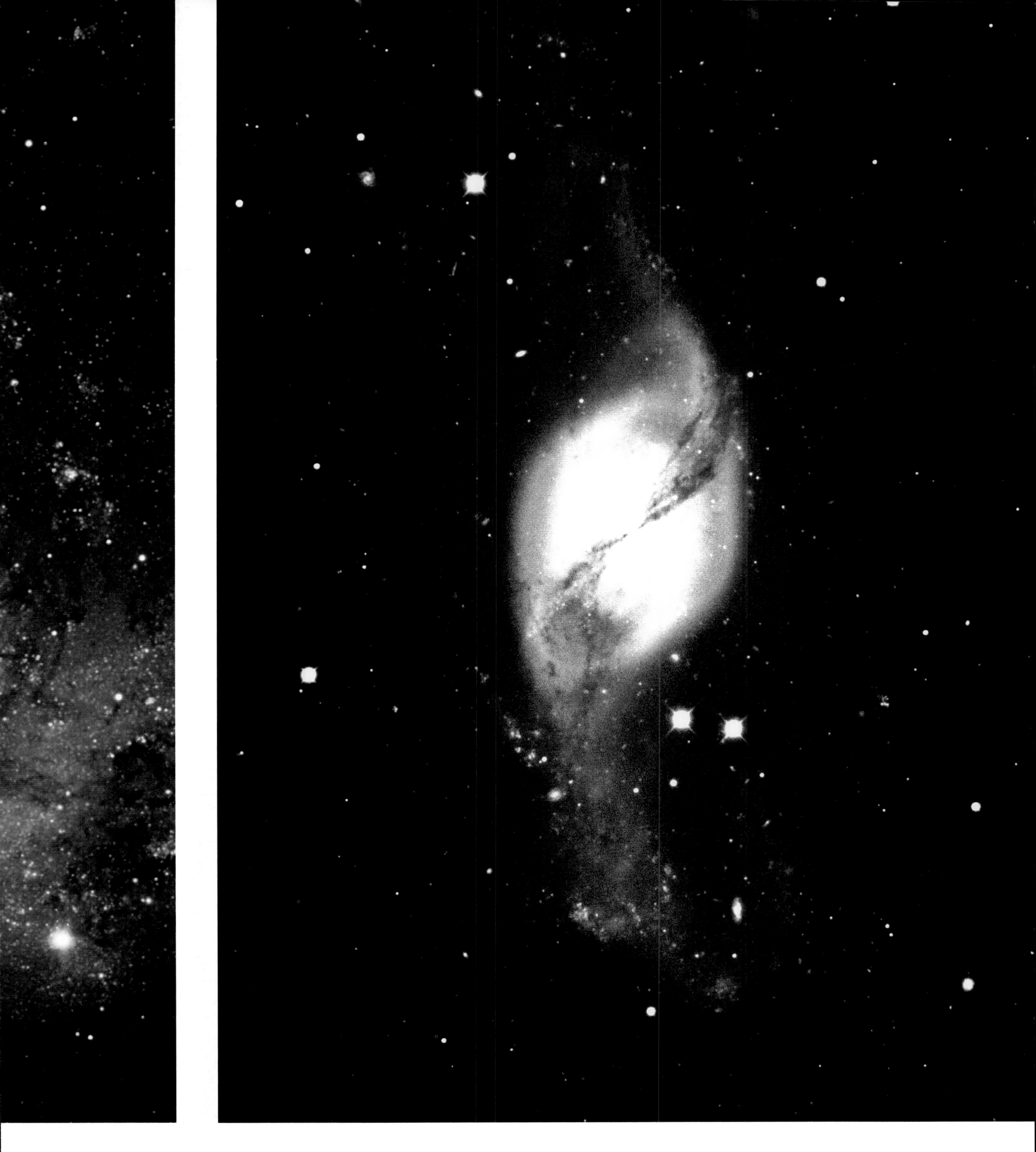

Peculiar Galaxy NGC 3718

IN COMPLETE CONTRAST, NGC 3718 is poor in gas and is forming few young stars. But this lenticular (lens-shaped) galaxy compensates by having a glowing, active core—probably driven by a supermassive black hole. An impressive dark lane of dust crosses right in front of its violent nucleus. Pioneering research by British astronomer Linda Sparke (based in Wisconsin), has revealed that the disk of NGC 3718 is warped through an angle of 90 degrees by the gravity of dark matter in the galaxy's halo. Eventually, the warp will develop into a ring of gas at a right angle to the galaxy's disk.

NGC 1232

THIS GLORIOUS GALAXY WAS PHOTOGRAPHED BY ANTU, one of the four huge 27-foot/8.2-m telescopes at the VLT in Chile. NGC 1232 is a so-called grand design spiral: its architecture is symmetric, beautifully proportioned, and utterly classical. Located about 100 million light-years away in the constellation of Eridanus, the galaxy's spiral arms are delineated by hot, glowing nebulae of hydrogen gas in which new stars are being born. The small galaxy on the left of NGC 1232 is not associated and lies in the depths of the cosmic background.

Spiral Galaxy NGC 7331

GALAXY NGC 7331 IS ONE OF THE CHOSEN FEW. It has been selected as one of eighteen by the team who run the Hubble Space Telescope to measure precise distances in the universe. The galaxy, 49 million light-years away, contains Cepheid variable stars. These swell and shrink according to their size—a bigger star takes longer to vary than a smaller one. By comparing these stars with Cepheids in the Milky Way, whose distances are known, astronomers can then figure out how far away other galaxies are. In this way, we can find out the scale—and the age—of the universe.

Spiral Galaxy NGC 4013 ▷

A DUST LANE 500 LIGHT-YEARS THICK appears to cut the galaxy NGC 4013 in half. This image—captured by the Hubble Space Telescope with a 1.7 hour–long exposure—reveals about 35,000 light-years of this spiral galaxy, toward its outer edge. The galaxy lies 55 million light-years away from us in the constellation of Ursa Major. A brilliant star shines out like a glittering diamond against the dust lane, but it is a foreground star in our Milky Way, and not a member of NGC 4013.

Sombrero Galaxy M104

SO EVOCATIVE OF A CELESTIAL SOMBRERO HAT, this aptly named—and most photogenic—galaxy lies about 30 million light-years away in the constellation of Virgo. The "brim" of this giant cosmic hat is etched out by layers of sooty dust lanes, poised to collapse and create new stars. At the heart of this tightly wound spiral galaxy is a huge black hole, weighing in at a billion Suns. Two thousand globular clusters swarm in the galaxy's massive halo—more than ten times the number that surround the Milky Way. This image, captured by the Hubble Space Telescope, is one of the most detailed it has ever taken.

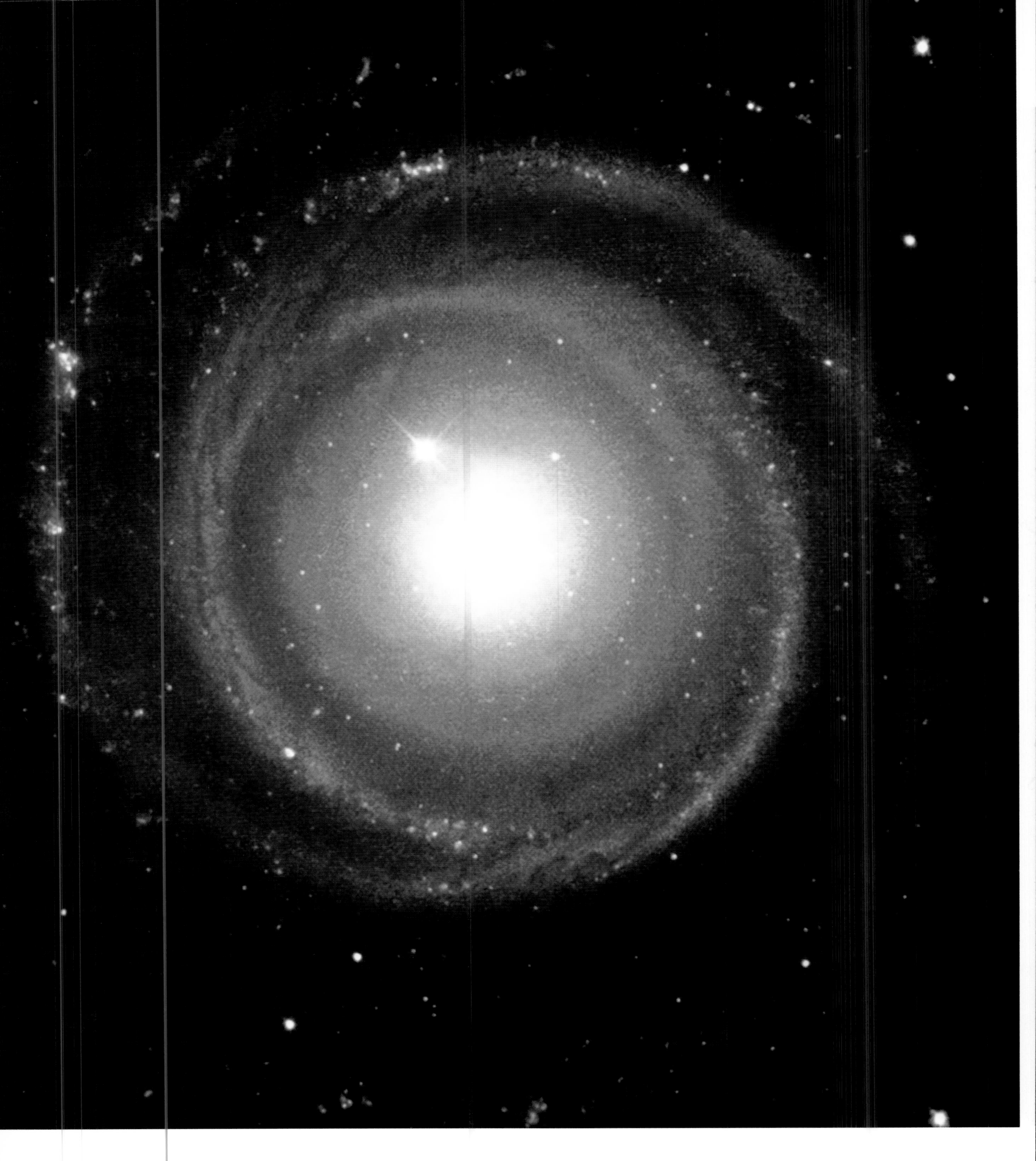

Galaxy NGC 4622

ONE HUNDRED MILLION LIGHT-YEARS AWAY, in the constellation of Centaurus, NGC 4622 is a cosmic enigma. Most astronomers would look at this Hubble Space Telescope image of the galaxy and assume—from its outer bluish spiral arms—that it is rotating counterclockwise. But look at the reddish inner arms, and it seems to be spinning clockwise. The latest data, however, seem to show that it is rotating clockwise: its outer spiral arms are opening outward in the direction of rotation. Researchers believe that a past collision with a smaller companion galaxy could have led to the bizarre appearance of this galaxy.

Galaxy M109

A BARRED SPIRAL GALAXY about 50 million light-years away in the constellation of Ursa Major, M109 was logged by the eighteenth-century French astronomer Charles Messier as one of his "vermin of the skies." In his quest to track down comets, Messier cataloged over one hundred misleading, cometlike objects, never realizing that his list would become his legacy to astronomy. M109 is a member of the Ursa Major Cloud, a loose agglomeration of galaxies. Its spiral arms are dotted with star-forming regions.

◁ Spiral Galaxy NGC 4603

NASA'S HUBBLE SPACE TELESCOPE photographed this beautiful spiral galaxy in Centaurus in 1999. NGC 4603 lies 108 million light-years away—so far away that even Hubble strains its eyes to pick out its individual stars. This is the most distant galaxy yet found in which Cepheid variable stars have been detected. Astronomers on the Space Telescope team have been monitoring their well-known brightness changes and have succeeded in pinning down an accurate distance for this elegant spiral.

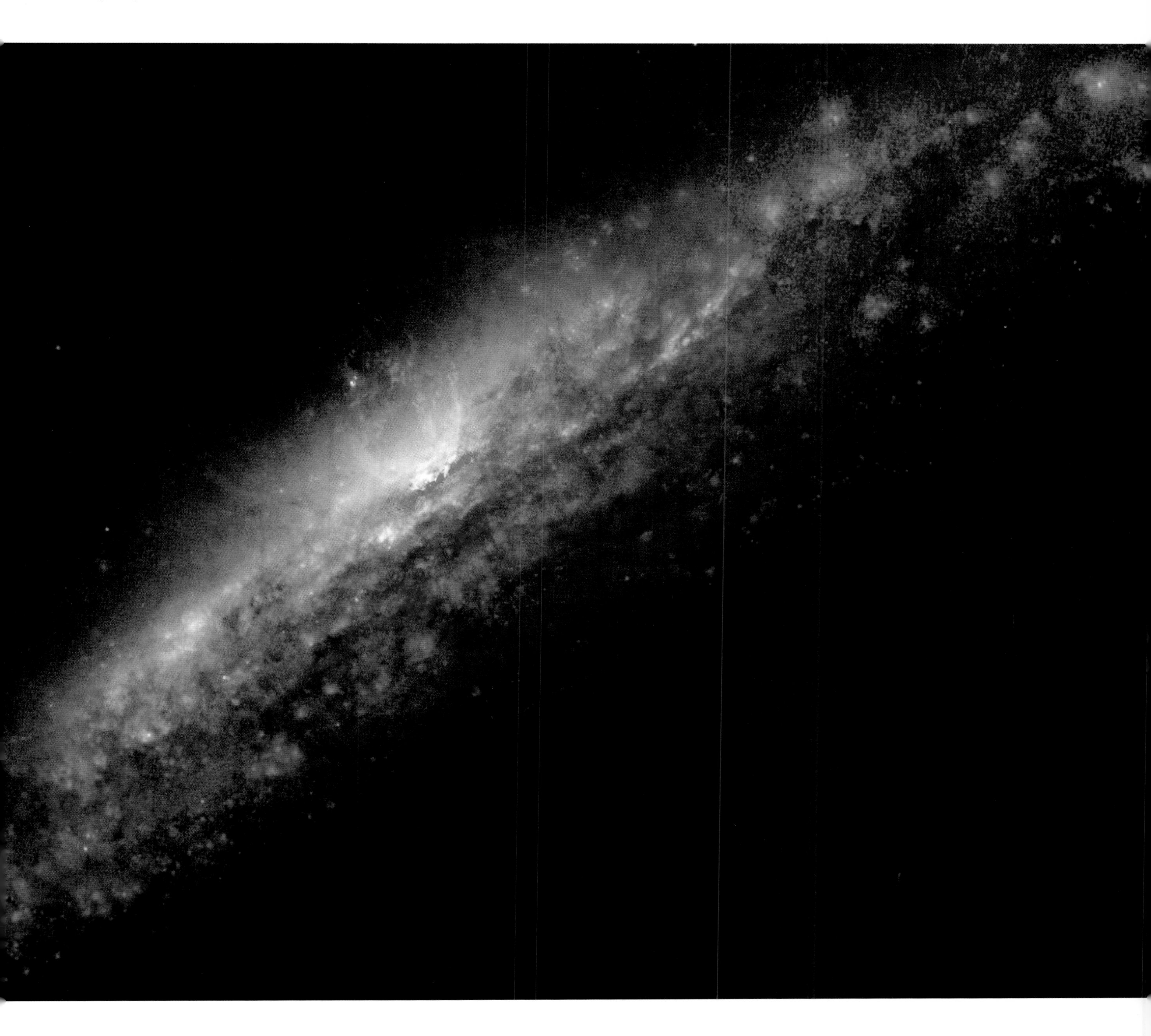

Spiral Galaxy NGC 3079

THIS STUNNING FALSE-COLOR IMAGE from the Hubble Space Telescope showcases the galaxy NGC 3079, located 50 million light-years away in the constellation of Ursa Major. This spiral, seen from the edge, spans some 70,000 light-years. Its most sensational feature is the exploding bubble at its core. This cosmic cauldron has pillars of gas towering 2,000 light-years above it, streaming away at speeds of 3,700,000 miles per hour/6,000,000 kph. Astronomers believe that the activity is due to a recent burst of star formation in which violent stellar winds are driving the gas outward.

NGC 2787

GALAXIES ARE NOT ALWAYS THE EPITOME OF GLAMOUR—especially when they grow old. This is true of NGC 2787, a lenticular galaxy about 24 million light-years away in the constellation of Ursa Major. Its bland appearance is enhanced by narrow lanes of sooty dust, which wrap themselves around its body of old red stars. The galaxy is circled by at least twelve old globular clusters, which may represent some of the leftover building blocks of this aging collection of stars.

Halo around NGC 4631

IN CONTRAST, HERE IS A GALAXY BURSTING WITH YOUTH: NGC 4631—which lies about 25 million light-years away from us—is a galaxy very like our Milky Way. This false color image is the result of a combined effort between the Chandra X-Ray Observatory (which detects hot gas, shown here in blue) and the Hubble (which looks at stars, shown here in red). It reveals a maelstrom of star formation. The bubbles in the galaxy's disk are created by clusters of young, massive stars. Surrounding NGC 4631 is an enormous halo (picked out by Chandra), which almost certainly contains dark matter.

Barred Spiral Galaxy NGC 1300

IN JANUARY 2005, excited astronomers at the meeting of the American Astronomical Society in San Diego, unveiled this awesome photograph of the barred Spiral Galaxy NGC 1300. Presenting it to the audience as an image measuring 4 x 8 feet/1.2 x 2.4 m across, the scientists pointed out that this is one of the most detailed pictures of a galaxy ever captured. Sixty-nine million light-years away, in the constellation of Eridanus, NGC 1300 sports blue and red supergiant stars, star clusters, star-forming regions, and delicate dust lanes in the central nucleus.

Spiral Galaxy M66

VISIBLE IN A SMALL TELESCOPE, this barred spiral galaxy in Leo lies about 35 million light-years away. It makes up part of a triplet of galaxies, one of which—M65—may have distorted the arms of this galaxy because of its gravitational pull. The "bar" at the center of M66 is a result of its nucleus rotating as a single unit. M66 is active in making stars. However, many young, massive stars explode as supernovas. There was a spectacular supernova in M66 in 1989. Since these exploding stars reach much the same brightness, they can help astronomers calibrate the scale of the universe.

Spiral Galaxy NGC 7424

A GALAXY THAT IS TRULY EYE-CATCHING, NGC 7424 is about 40 million light-years away in the southern hemisphere constellation of Grus. It is probably the galaxy in the sky most similar to our own Milky Way, having a central small nuclear bar and open spiral arms. At 100,000 light-years across, NGC 7424 is almost identical in size to our galaxy and shares with it a scattering of beautiful nebulae and star clusters. This photograph was taken by Melipal, one of the 27-foot-/8.2-m-diameter telescopes at Europe's VLT Observatory in Chile. Melipal means "Southern Cross" in the local dialect, Mapuche.

Spiral Galaxy NGC 1350

THIS GORGEOUS ISLAND OF STARS lies about 85 million light-years away in the southern constellation of Fornax. Dotted with young, blue star clusters, the slender arms of NGC 1350 curve in a circle around its nucleus—giving the galaxy the appearance of a limpid cosmic eye. At 130,000 light-years across, NGC 1350 is larger than our Milky Way. Yet its delicacy allows the light from other galaxies to be seen through its spiral arms. This image was captured by Kueyen—one of the VLT's telescopes, whose name means "moon" in Mapuche.

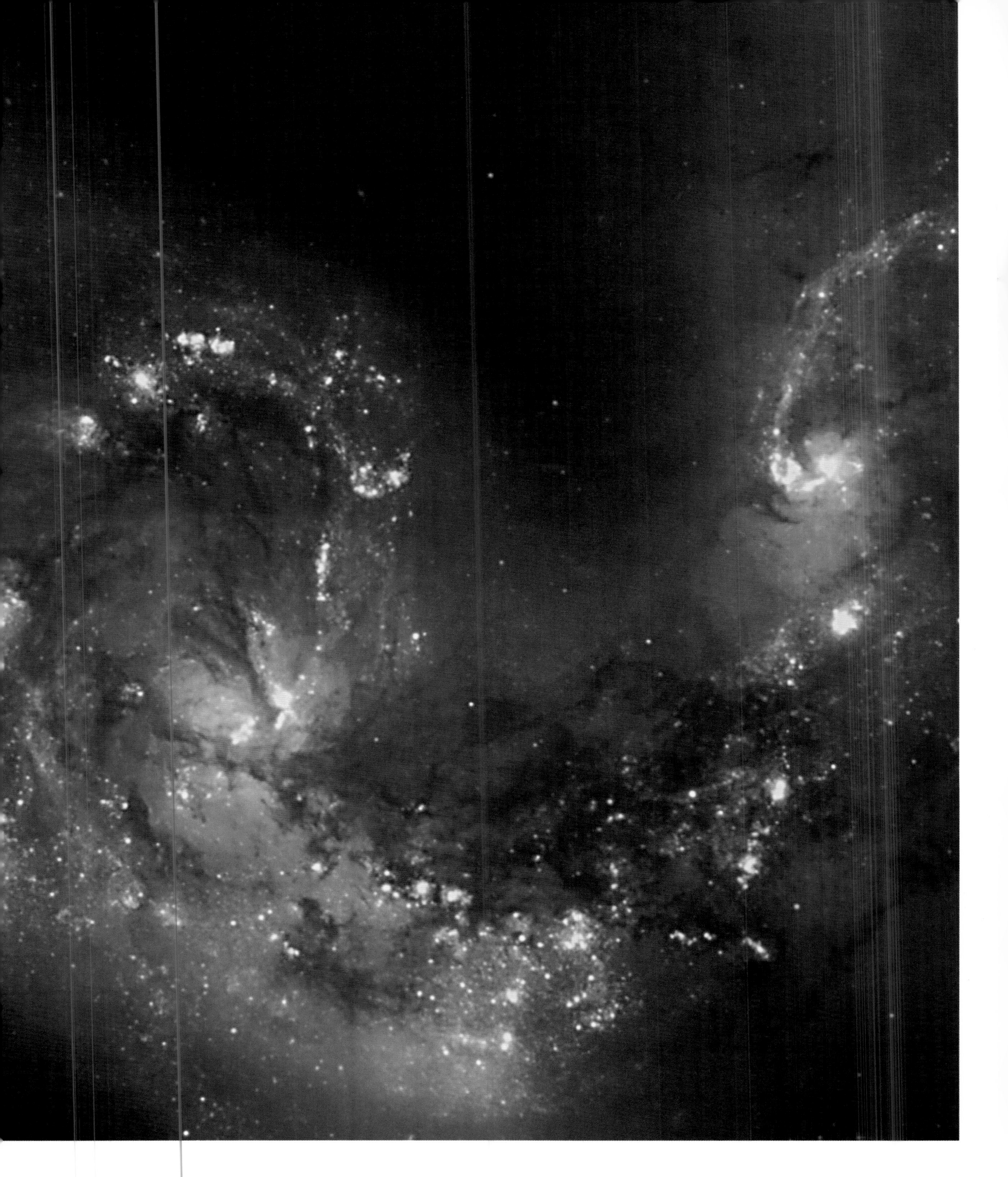

Antennae interacting galaxies

WHEN GALAXIES GET UP CLOSE AND PERSONAL with one another, they can wreak havoc. The Antennae are two galaxies that have done just that. Named after the arcing, insectlike "antennae" of gas that have arisen from their collision, NGC 4038 and NGC 4039 are a cosmic traffic accident writ large. A distance of 63 million light-years away, the nuclei of the two galaxies are just 30,000 light-years apart. Astronomers using the Hubble Space Telescope are now sorting through the wreckage—and their work has led to the discovery of over 1,000 bright young star clusters triggered by the collision.

Interacting galaxies NGC 6872 and IC 4970

MEASURING OVER 700,000 LIGHT-YEARS FROM TAIL TO TAIL—seven times the diameter of the Milky Way—NGC 6872 is one of the most elongated spiral galaxies known. It is performing a cosmic ballet with the small galaxy IC 4970 (just above center). The smaller galaxy's gravity has caused this disruption, flinging tendrils of star-forming gas from the spiral into space. The astronomers who captured this image—with the telescope Antu ("sun") at the VLT in Chile—have pointed out that light from this pair of galaxies set off on its journey 300 million years ago, long before the dinosaurs existed on the Earth.

Interacting galaxies NGC 2207 and IC 2163

FIVE BILLION YEARS HENCE, this fate will befall the Milky Way and the Andromeda Galaxy. The two star systems—which look so similar to our galaxy and its larger neighbor—are closing in on one another and will eventually merge. NC 2207 (left) and IC 2163 (right) are already embroiled in a cosmic embrace. Their most recent encounter took place just 40 million years ago—a short moment ago in cosmic time. The two galaxies will slowly pull each other apart, creating violent tides, bursts of star formation, and cast-off stars. Eventually, the larger galaxy will devour its smaller prey.

Whirlpool Galaxy M51

THE WHIRLPOOL GALAXY IS A CLASSIC. The Irish astronomer Lord Rosse—who discovered "Spiral Nebulas" (as they were then called) in the mid-nineteenth century—proclaimed that "this was the most conspicuous of the spiral class." Its small companion galaxy, NGC 5195, had a narrow brush with the Whirlpool 70 million years ago. Gravitational tides between the two galaxies created the fabulously symmetrical structure that we see in this beautiful spiral today. The gas drawn out by the encounter no longer links the two galaxies: NGC 5195 just happens to lie in the line of sight of the stream.

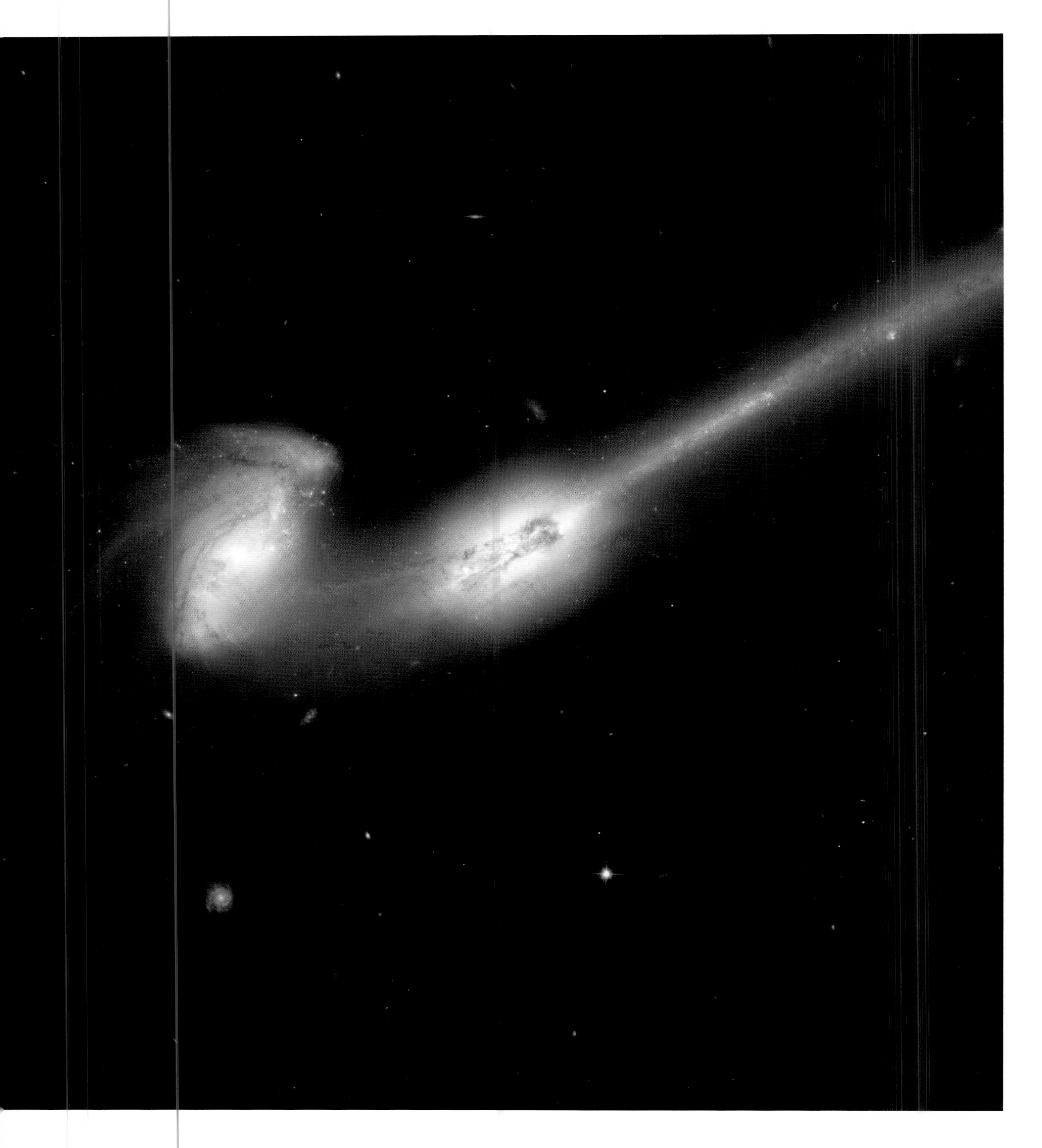

The Mice colliding galaxies

NAMED FOR THEIR LONG TAILS, THE MICE are a sensational pair of interacting galaxies. Arching millions of light-years into space, their tidal streamers—made up of vast amounts of gas and stars thrown off by the encounter—strew themselves across the cosmos. Formerly two spiral galaxies, this currently brilliant duo will merge to become a bland elliptical galaxy, bereft of gas, in a few million years time. In the meantime, these two active spirals will undergo an orgy of star formation: the Mice are already resplendent in hot, young blue stars. This pair of galaxies lies 300 million light-years away in the constellation of Coma Berenices.

The Tadpole colliding galaxies

IN THIS STUNNING VISTA CAPTURED BY THE HUBBLE SPACE TELESCOPE, distant galaxies form a backdrop for the disrupted remains of the Tadpole Galaxy. Named for its shape, this cosmic mini-frog lies 420 million light-years away in the constellation of Draco. The Tadpole's tail is a result of an interaction between the main galaxy (Arp 188) and a small galaxy about 300,000 light-years behind (seen through its spiral arms, upper left). Following its terrestrial namesake, the Tadpole Galaxy will lose its tail as it grows older, creating baby star clusters out of the torn-out gas born out of the collision.

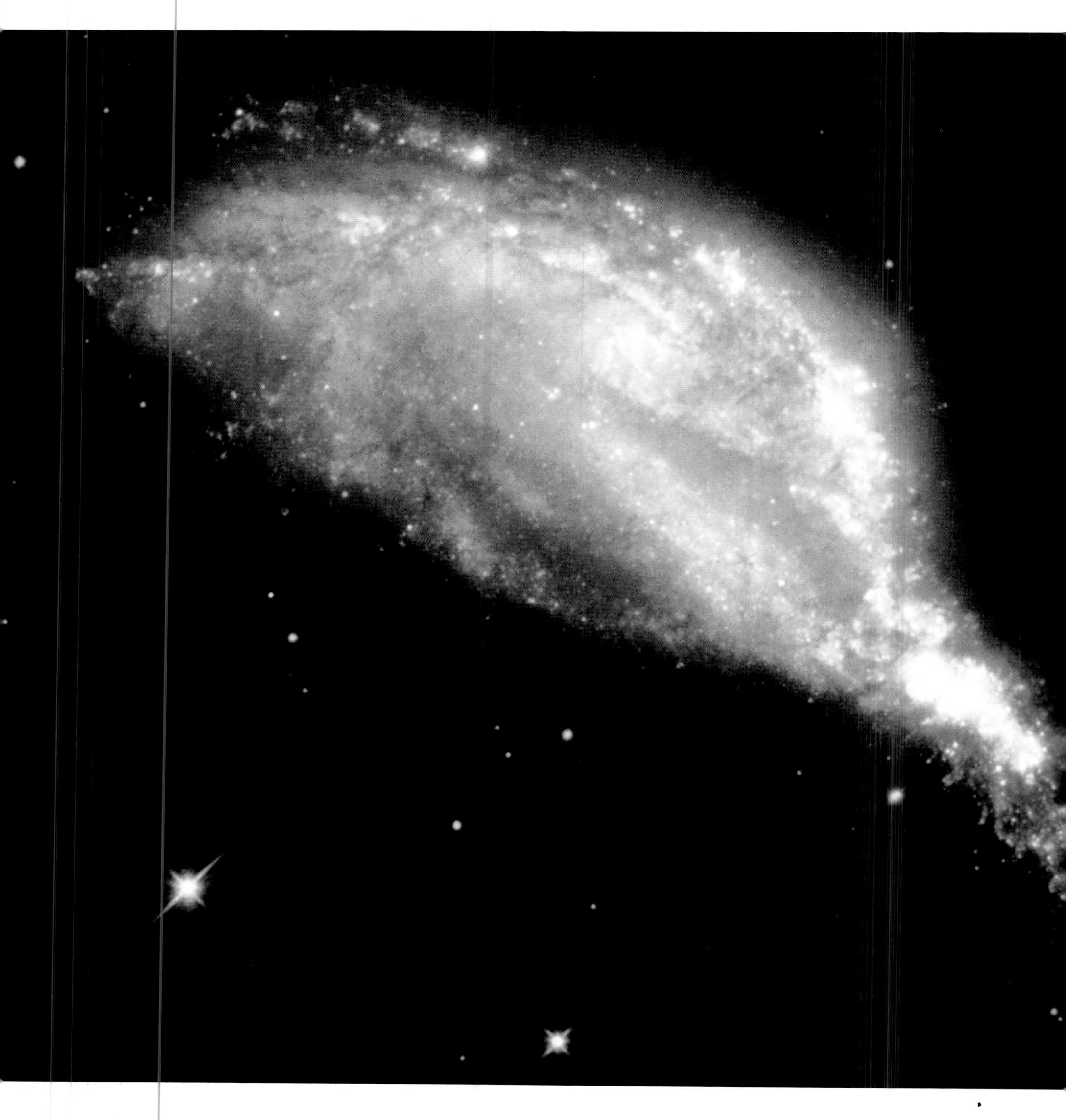

Interacting galaxies NGC 6745

THE NGC 6745 SYSTEM GRAPHICALLY DEMONSTRATES the result of two galaxies that have been engaged in a cosmic tussle for hundreds of millions of years. In the past, the larger galaxy used to be a conventional spiral like our Milky Way, but the gravity of its companion (lower right), has maimed and distorted it. The interaction has also led to an outburst of star formation (brilliant white cloud, right). These two galaxies are 200 million light-years away and span a distance of 80,000 light-years.

Hubble image of colliding galaxies NGC 1275

AN INTERGALACTIC CLASH OF THE TITANS: in the NGC 1275 system, two mighty galaxies slice through one another. At top left, an elliptical galaxy collides with a dusty spiral (center). These extraterrestrial encounters result in cosmic mayhem, leading to violent outbursts of star formation. These two galaxies, seen in a composite image taken by the Hubble Space Telescope in 1995 and 2001, lie in the Perseus Cluster of galaxies. The cluster is 200 million light-years away, and the galaxies each measure 50,000 light-years across.

Ring Galaxy AM 0644-741 ▷

HOW COULD A GALAXY BE SHAPED LIKE A RING? From a head-on collision between a large spiral galaxy (top right), and a smaller intruder. In this case, the intruder has moved on, leaving behind it a cosmic annulus 150,000 light-years across. The passage of the rogue galaxy compresses the gas in its victim, leading to an orgy of star formation. The waves of star birth move out from the center of the galaxy, just like ripples spread across a pond. AM 0644-741's ring is pregnant with huge nebulae and star clusters. The galaxy lies about 300 million light-years away, in the constellation of Dorado.

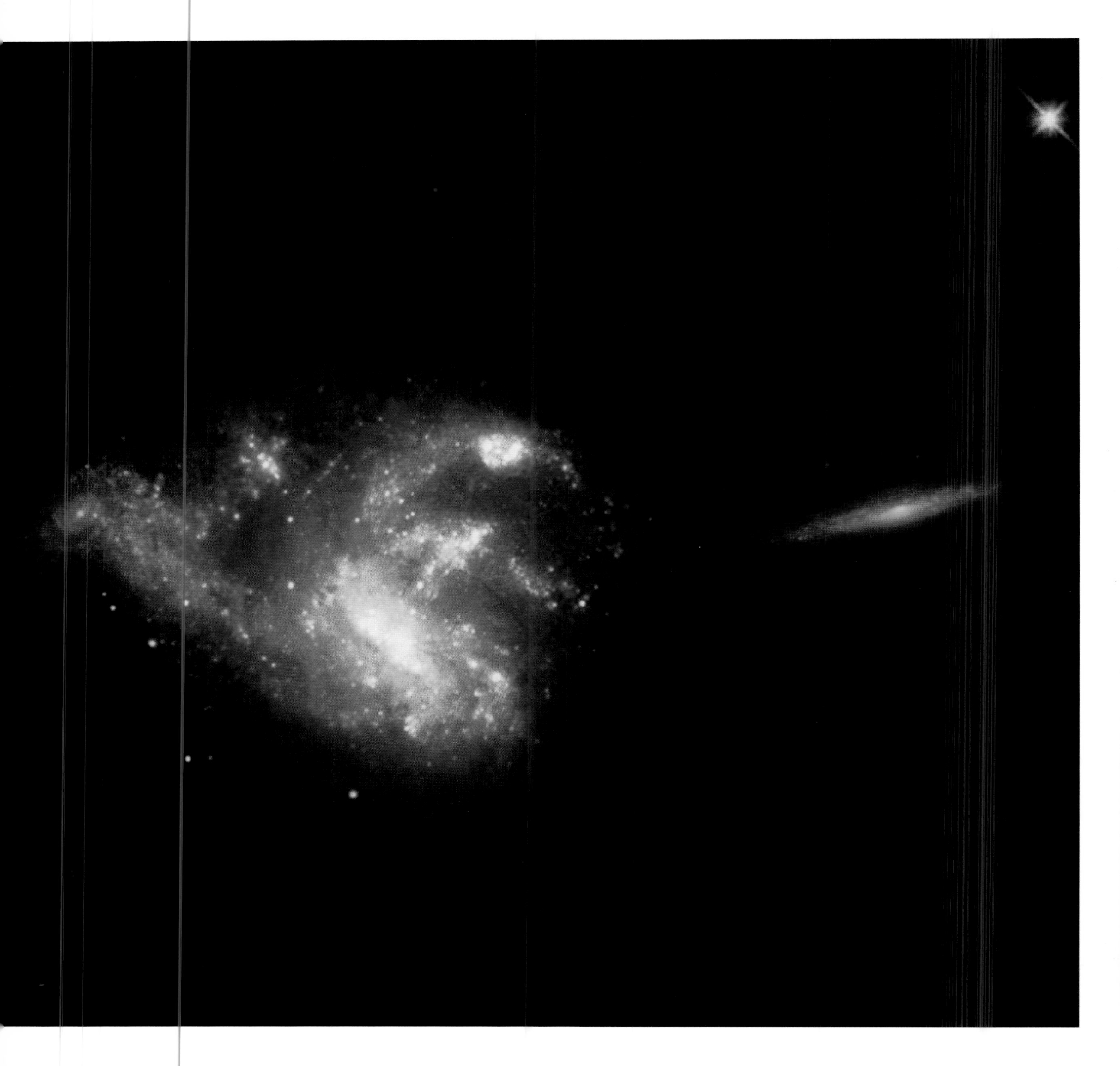

Irregular Galaxy NGC 7673

ACCORDING TO ASTRONOMERS from the European Space Agency and the Hubble Space Telescope (which captured this image), NGC 7673 is a "hyperactive galaxy." "It is ablaze with the light from millions of new stars," observes Nicole Homeier of Johns Hopkins University, "What has triggered this enormous burst of star formation, and how will this galaxy evolve in the future?" The brilliant young star clusters in the galaxy are possibly the result of a near-miss collision with a nearby galaxy. In this image the other two galaxies (left and right) are far away in the cosmic background. NGC 7673 itself lies about 150 million light-years away in the constellation of Pegasus.

Interacting galaxies NGC 6769, 6770, 6771

THIS EXQUISITE COSMIC BALLET between three galaxies is taking place in the southern constellation of Pavo—the Peacock—190 million light-years away. The two large spiral galaxies, NGC 6769 and 6770, are of equal brightness and are tangled in a mutual embrace. NGC 6771 (below) is an outside witness, although astronomers believe it is part of the interaction. Although violent, encounters like this lead to cosmic baby booms, as testified by the blue light from young stars shining from these star systems. This image was taken by the VLT in Chile to commemorate its fifth anniversary on April 1, 2004.

Colliding galaxies NGC 1531 and 1532

FIFTY-FIVE MILLION LIGHT-YEARS AWAY in the southern constellation of Eridanus, the galaxies NGC 1532 (the large spiral, seen from its side) and its smaller companion (NGC 1531) are getting their act together in a major way. This photograph was taken by the 26.5-foot/8.1-m Gemini telescope in Chile (its twin is on Hawaii) to celebrate the launch of the observatory's image gallery in December 2004. When galaxies interact, there are celestial fireworks—riots of star formation and incandescent nebulae—but the stars themselves seldom collide.

Andromeda Galaxy

EIGHT TIMES THE DIAMETER OF THE FULL MOON in our skies, the Andromeda Galaxy—at a distance of 2.9 million light-years—is easily visible as a hazy oval to the unaided eye in dark locations. In AD 964, the Persian astronomer Al-Sufi recorded it as "a little cloud." But this monster of a galaxy is twice as large as the Milky Way, containing 400,000 million stars. It is almost a mirror-image of our own home in space, with its spiral arms and two companion galaxies.

Andromeda Galaxy nucleus

THE HEART OF OUR NEXT-DOOR SPIRAL NEIGHBOR, the Andromeda Galaxy. Like the nucleus of our Milky Way, the center of Andromeda is an egg-shaped ball of aging red and yellow stars. Ironically, we can study its very core in more detail than we can examine the downtown regions of our own galaxy, because we can see the center of Andromeda unobscured by dust in the line of sight. The galaxy appears to have two hot spots at its heart. Astronomers speculate that these may be two satellite galaxies swallowed by Andromeda—and that one is now devouring the other.

Pinwheel Galaxy M33

YOU ARE LOOKING AT THE MOST DISTANT OBJECT visible to the unaided eye. Amateur astronomers have reported seeing this faint galaxy, three million light-years away, on extremely dark, moonless nights. M33—a ragged spiral half the size of the Milky Way—is, like the Andromeda Galaxy, a member of our "Local Group." This tiny cluster of around fifty largely unspectacular galaxies is our galactic cosmic community. M33—in the constellation Triangulum—is physically very close in space to the neighboring Andromeda Galaxy. The views of one another from their respective homes must be sensational.

Triangulum (Pinwheel) Galaxy

M33, THE PINWHEEL GALAXY, is alive with nurseries of star formation. Baby stars are incubating in the pink nebulae dotted around the straggly spiral arms, and their fierce radiation is exciting the natal gas to glow. The biggest Nebula, NGC 604, is on the bottom left. This enormous cosmic star factory—1,500 light-years across—was discovered by William Herschel (who found the planet Uranus) on September 11, 1784. Very similar to the vast Tarantula Nebula in our galaxy's neighboring Large Magellanic Cloud, NGC 604 has already given birth to two hundred stars. They are stellar giants: their masses range from fifteen to sixty times that of our Sun.

Centaurus A

TWELVE MILLION LIGHT-YEARS DISTANT, the galaxy Centaurus A is a cosmic giant. It contains a million, million stars, which makes it a striking object visible from the southern hemisphere—it is easily seen through binoculars. This huge elliptical galaxy, however, is a cannibal. The tortured rift through its center is the remains of a spiral galaxy that collided with Centaurus A 100 million years ago—and which the larger galaxy is currently digesting. Material from the spiral is providing a feeding frenzy for the black hole at the core of Centaurus A, which weighs in at 200 million times the mass of the Sun.

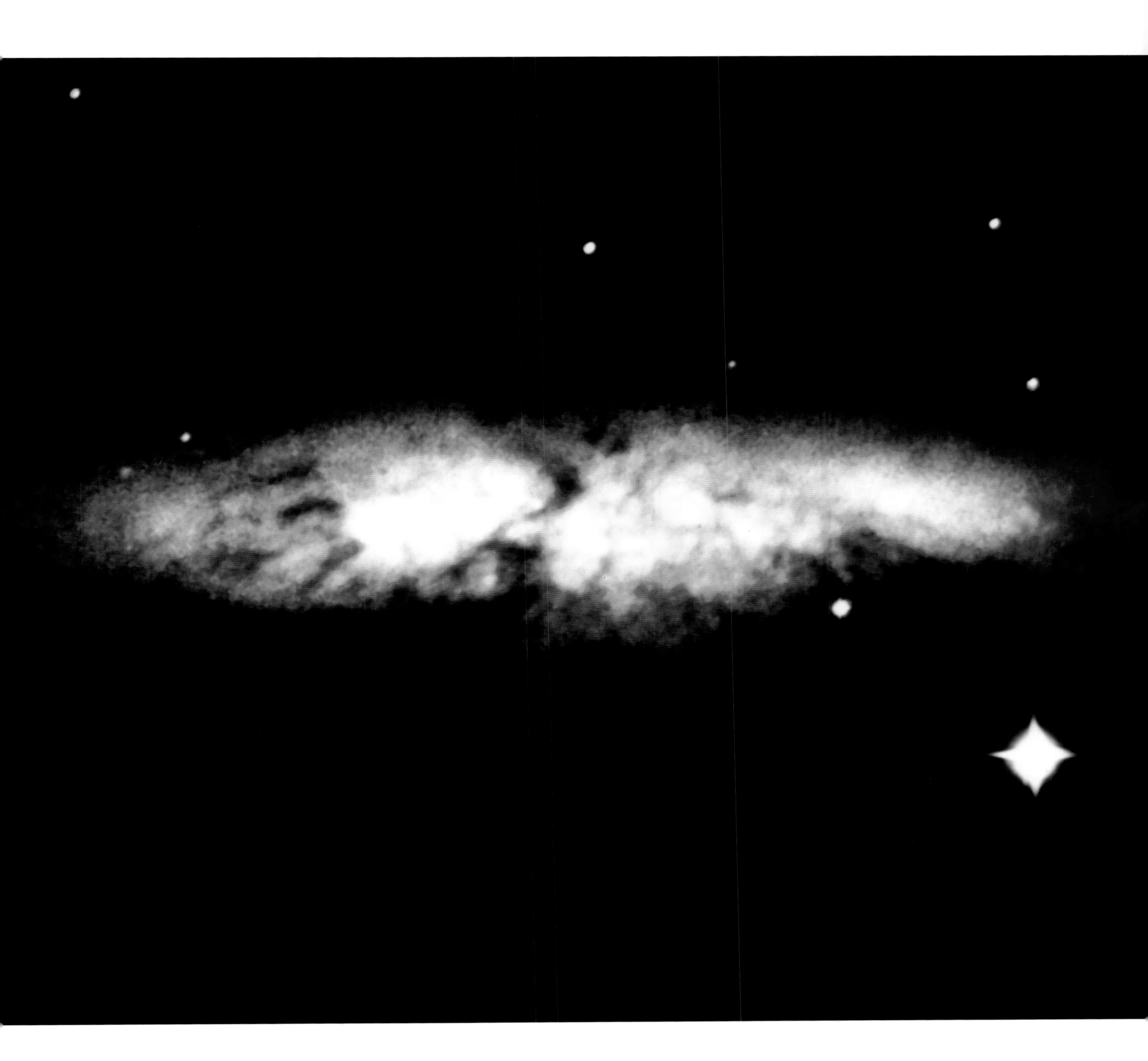

Galaxy M82

ALSO TWELVE MILLION LIGHT-YEARS AWAY, M82 lies in a different direction from Centaurus A, in the northern constellation of Ursa Major. Basically, the "Cigar Galaxy" (as it is nicknamed), is a mess—and astronomers have been poring over its structure for well over one hundred years to figure out what makes it tick. The latest thinking is that its urbane spiral neighbor, M81, made a recent flyby and triggered a violent outburst of star formation in M82's nuclear regions. The resulting "superwinds" from the young stars are churning up the galaxy's heart.

"BLACK HOLE = NAUGHTY WORD," SCRAWLED AN ASTRONOMER—reminiscent of Walt Whitman's "learn'd astronomer"—on an old-fashioned blackboard during a rather tedious address at an international conference in 1970.

Alas for him, the writing was already on the wall. In 1966, a young research student at Cambridge—Jocelyn Bell (now Professor Bell-Burnell)—discovered a new type of object in the universe. It was a neutron star: the supercompressed core of a star that had exploded as a supernova. The material in this beast is squeezed so tightly that a pinhead of its material would weigh as much as a supertanker. Even light, traveling at the cosmic speed limit—186,000 miles per second/300,000 km/s—has problems escaping a gravitational field as powerful as this.

It was inevitable that something even more extreme would come up. And in the very year that the astronomer chalked up his fateful words, an American satellite, Uhuru, detected a powerful source of X-rays: one of the most energetic forms of radiation in space. In Cygnus X-1—the first black hole to be detected—an unseen object, heavier than ten Suns, was tearing matter off its companion star. The gas forms a swirling vortex—the accretion disk—around the hole itself. Friction makes the speeding gas extremely hot, causing the disk to glare brilliantly.

Black holes are a triumph of gravity. In the case of stars, they live on borrowed time. Gravity creates them; gravity destroys them. When massive stars die, they explode as supernovas. But they always leave behind their central core. If it weighs in at more than three Suns, not even superdense neutrons can hold it up against the force of gravity. Instead of turning into a neutron star, it collapses completely into a black hole. Light cannot escape the gravity of a black hole: it is black forever. And because nothing can travel faster than light, anything close to the hole becomes embroiled in its gravitational thrall. Black holes swallow up anything in their immediate surroundings.

You cannot see a black hole, but you can pinpoint one through its reckless lifestyle. Imagine a black cat and a white cat having an altercation in a coal cellar at night. You won't see the black cat, but the behavior of the white cat will reveal that there is something untoward going on.

In our galaxy alone, we know that there are dozens of black holes resulting from supernova explosions. But there are even bigger fish out there. The centers of many galaxies—including our own—harbor supermassive black holes. These cosmic behemoths weigh in at millions, or even billions of Suns. These are the result of the collapse of generations of stars. At their most extreme—in eruptive galaxies called quasars—they stir up huge explosions, creating jets of gas that travel close to the speed of light.

When matter falls into a black hole, where does it go? No one knows. Some theoreticians believe that it is crushed into a tiny point of infinite density at the black hole's heart, called a "singularity." Others maintain that black holes may be our understanding to a new geometry of space—and offer pathways to other universes.

9 BLACK HOLES AND QUASARS

M74: Chandra image

RED FOR DANGER . . . these celestial rubies warn of dozens of black holes lurking within a beautiful spiral galaxy. This city of stars, M74, is a twin to our Milky Way galaxy and lies 32 million light-years away. The Chandra satellite has taken a long hard look at M74 and discovered dozens of objects that are emitting X-rays. Each of these red dots marks the position of a black hole—the collapsed core of an old star. The radiation picked up by Chandra is the last shriek emitted by gas dragged from a still-living companion star and falling inexorably into the black hole.

Previous pages | Black hole

THE MOST MYSTERIOUS OBJECTS IN THE COSMOS, black holes are ravening monsters lurking in secret places. In this Hubble Space Telescope image of the center of the galaxy NGC 4261, a supermassive black hole gorges on a disk of dust and gas three hundred light-years across. The perpetrator, meanwhile, lies cloaked inside. Almost certainly, the victim of this cosmic cannibalism was a small galaxy that fell into the nucleus of its giant companion millions of years ago.

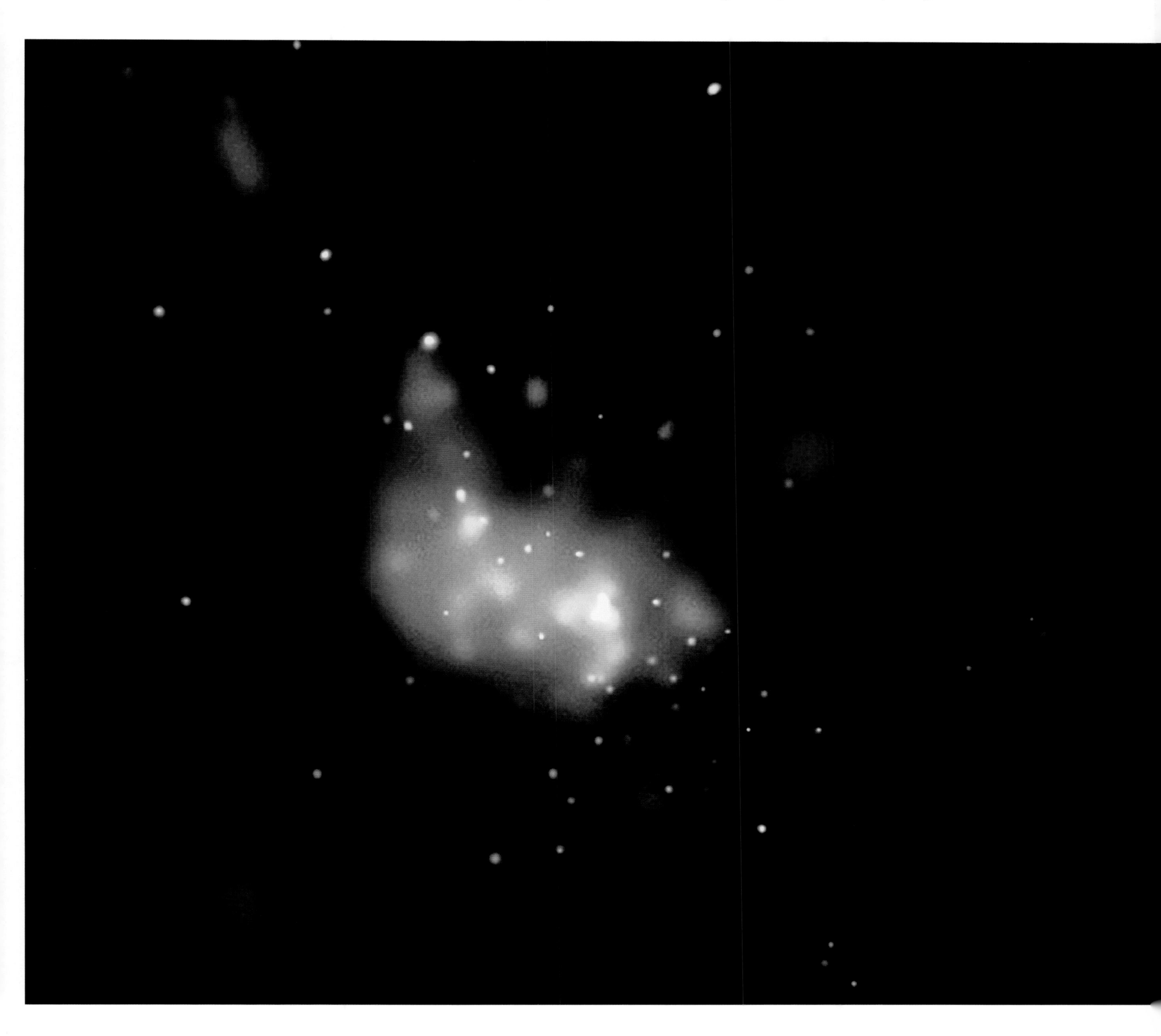

Galactic center

"WE HAVE TO FACE THE FACT that there is some outstanding peculiarity at the galactic center . . ." So warned the great Russian astrophysicist Josef Shklovskii back in the 1950s. He was well ahead of his time. Only recently have astronomers been able to confirm his prediction, with views like this from the Chandra satellite. The bright spot in the center marks the lair of the most monstrous black hole in our galaxy. Lying right at the heart of the Milky Way, it is a cloud of hot gas being swallowed by a giant black hole that weighs as much as 3.7 million Suns.

◁ Radio Galaxy NGC 1316

THE MAIN GALAXY IN THIS PICTURE looks nothing out of the ordinary—an elliptical ball of old stars, with a few wisps of dark dust running across it. But it is emitting an enormously powerful cacophony of radio waves. As a result, pioneering radio astronomers in the 1950s entitled it Fornax A: the major radio source in the southern constellation Fornax (the furnace). Astronomers now believe this radiation is powered by a supermassive black hole in the galaxy's core. It acts as a spinning dynamo, creating electric currents and magnetic fields that broadcast radio waves across the universe.

Seyfert Galaxy NGC 1365

A NEARBY NEIGHBOR OF FORNAX A—also lying around 60 million light-years away—is this huge spiral galaxy, NGC 1365, which is twice the size of our own Milky Way. And it has another claim to fame: at the very center of NGC 1365 lies a small brilliant point of light. This makes it a "Seyfert Galaxy," named for the young astronomer Carl Seyfert who first made a list of galaxies with starlike centers back in 1943—before being tragically killed in a car crash. Astronomers today believe the tiny brilliant core is a disk of superhot gas circling a massive black hole.

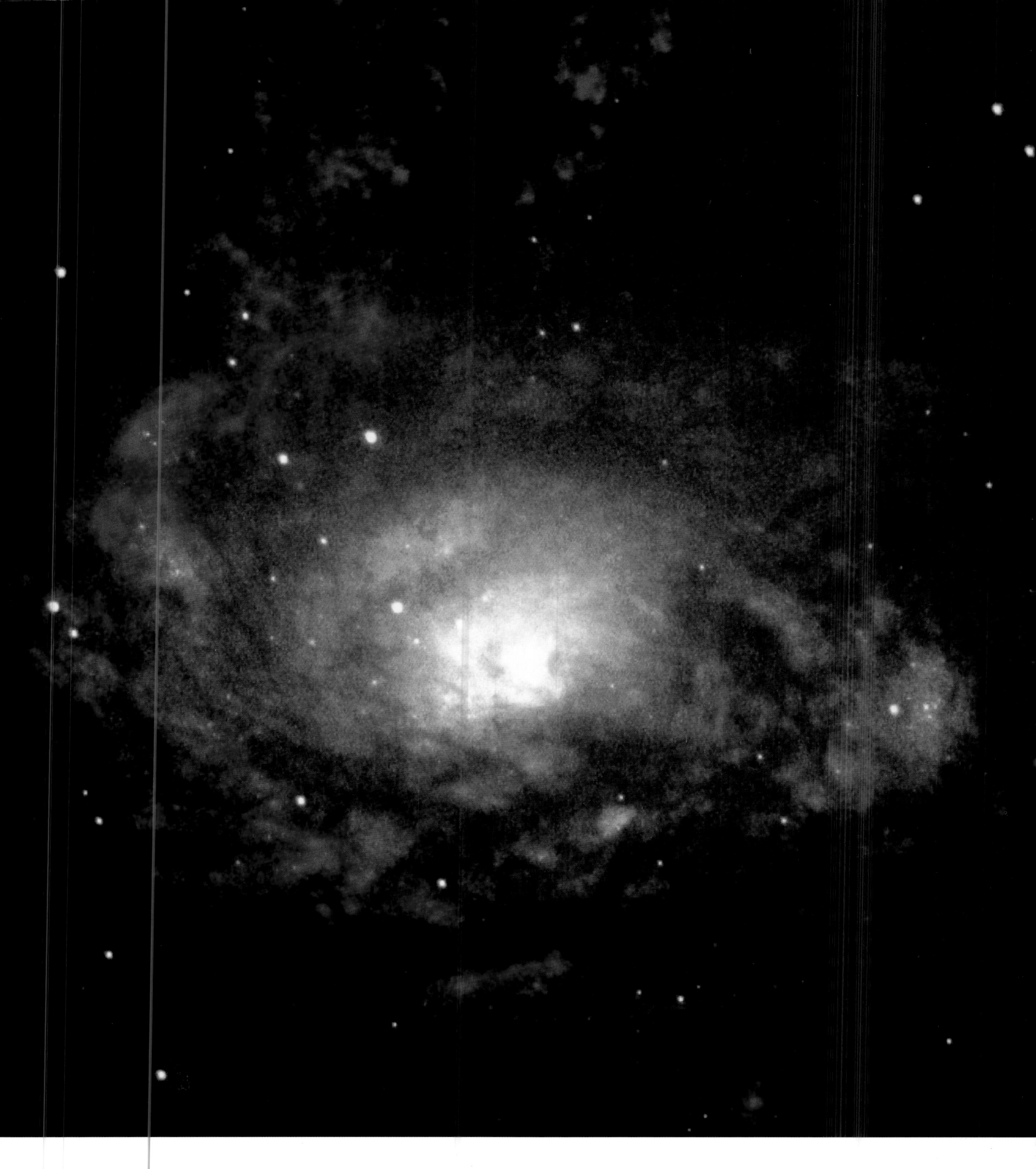

Circinus active galaxy

POWERFUL FORCES ARE EJECTING GAS from the core of the Circinus Spiral Galaxy. Plumes of pink hydrogen erupt upward in this view from the Hubble Space Telescope, while a ring of expanding gas is creating a rash of star birth that shows up in red. The activity originates from a supermassive black hole that is hidden at the galaxy's heart. Though Circinus is one of the nearest galaxies that is being disrupted by a black hole—just 15 million light-years away—astronomers overlooked it until recently because it is hidden behind dust clouds in the Milky Way.

Active Galaxy 0313-192

ONE BILLION LIGHT-YEARS AWAY lies a galaxy that looks much like our Milky Way, seen from its side, along with a companion that displays its spiral structure more clearly. The main sideways-view spiral is known only by its catalog number: 0313-192. And no one would pay it much attention were it not for two huge magnetic loops—seen in red here—that straddle the galaxy and broadcast powerful radio waves. The magnetism is undoubtedly generated by a hidden black hole at the galaxy's heart. But this galaxy represents a cosmic conundrum. Magnetic loops are almost always associated with round elliptical galaxies: why is this spiral galaxy sporting such strange appendages?

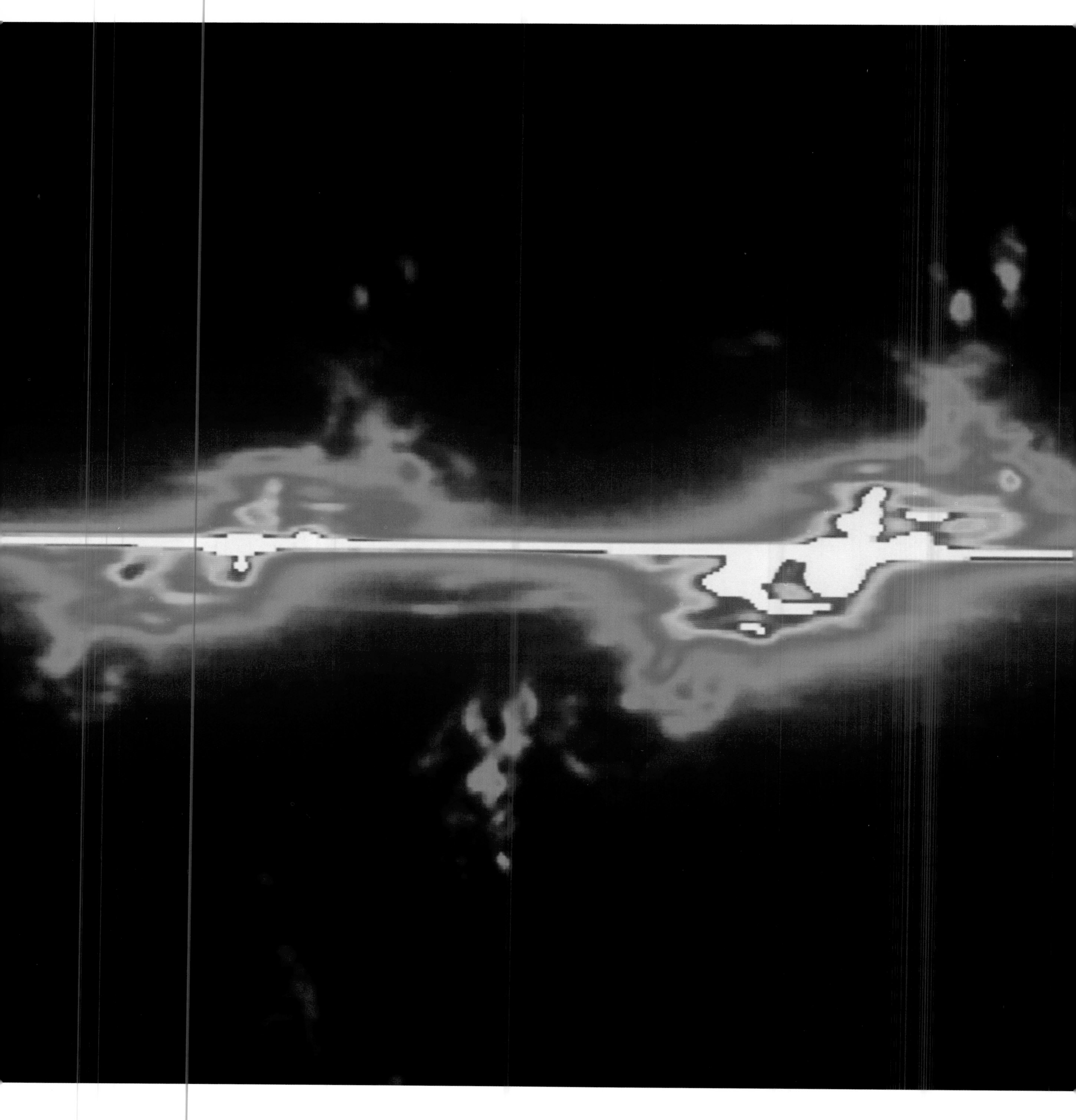

Seyfert Galaxy NGC 4151

THE HUBBLE SPACE TELESCOPE has here dissected the light from a galaxy to analyze the fireworks within. The object in question is NGC 4151, a prominent member of Carl Seyfert's original list of galaxies that harbor an unusual bright core. The light from the core is here spread out into a horizontal line. The two multicolored blobs, left and right, are formed by light that has come from oxygen atoms blasted out by the black hole at the galaxy's heart. By analyzing this image, astronomers have found that the black hole is ejecting gas at amazing speeds—over 100,000 miles per hour/161,000 kph.

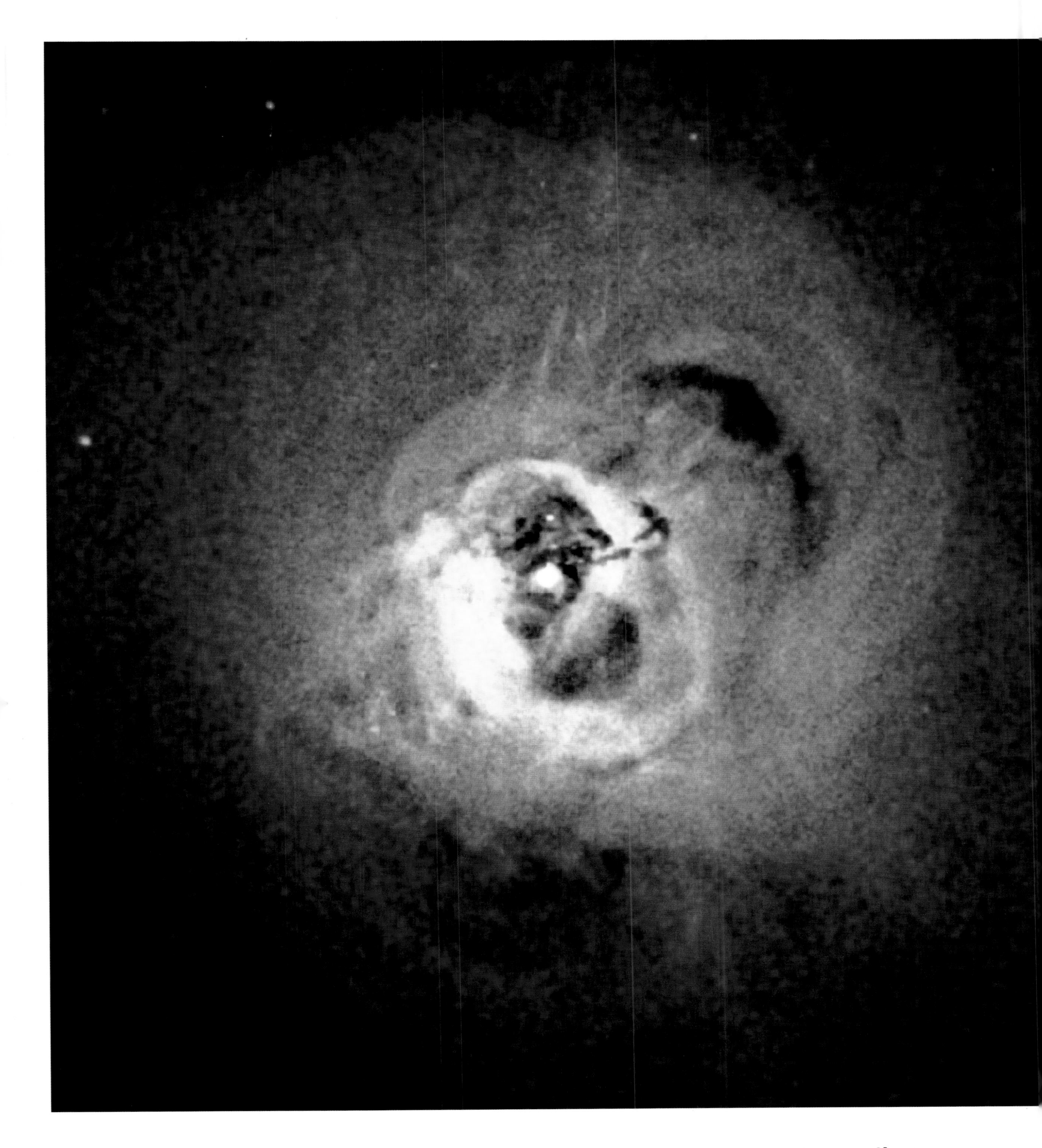

Perseus Cluster

A VAST CLOUD OF SUPERHOT GAS envelopes the galaxy Perseus A, as seen here in a view from the Chandra observatory. The dark greenish-blue strands are the remains of a galaxy that is being devoured by Perseus A. The central white spot marks Perseus A's central supermassive black hole. Explosions around the black hole are creating great blasts of noise, and the spreading sound waves can be seen in this image as ripples in the gas. The musical note played by Perseus A is B-flat, fifty-seven octaves below middle C. This frequency is over a million billion times deeper than the limits of human hearing,

Core of M87

LOCATED 60 MILLION LIGHT-YEARS AWAY, the Virgo Cluster is home to thousands of galaxies. But its king is surely M87—a true celestial giant—that lies at its heart. This massive elliptical galaxy contains 3 million, million stars. Center stage of this star system is a mighty black hole, weighing in at 2.6 billion Suns. In this Hubble Space Telescope image, the accretion disk surrounding the hole shines fiercely in brilliant white light. Yet this cosmic dynamo spans just the width of our solar system.

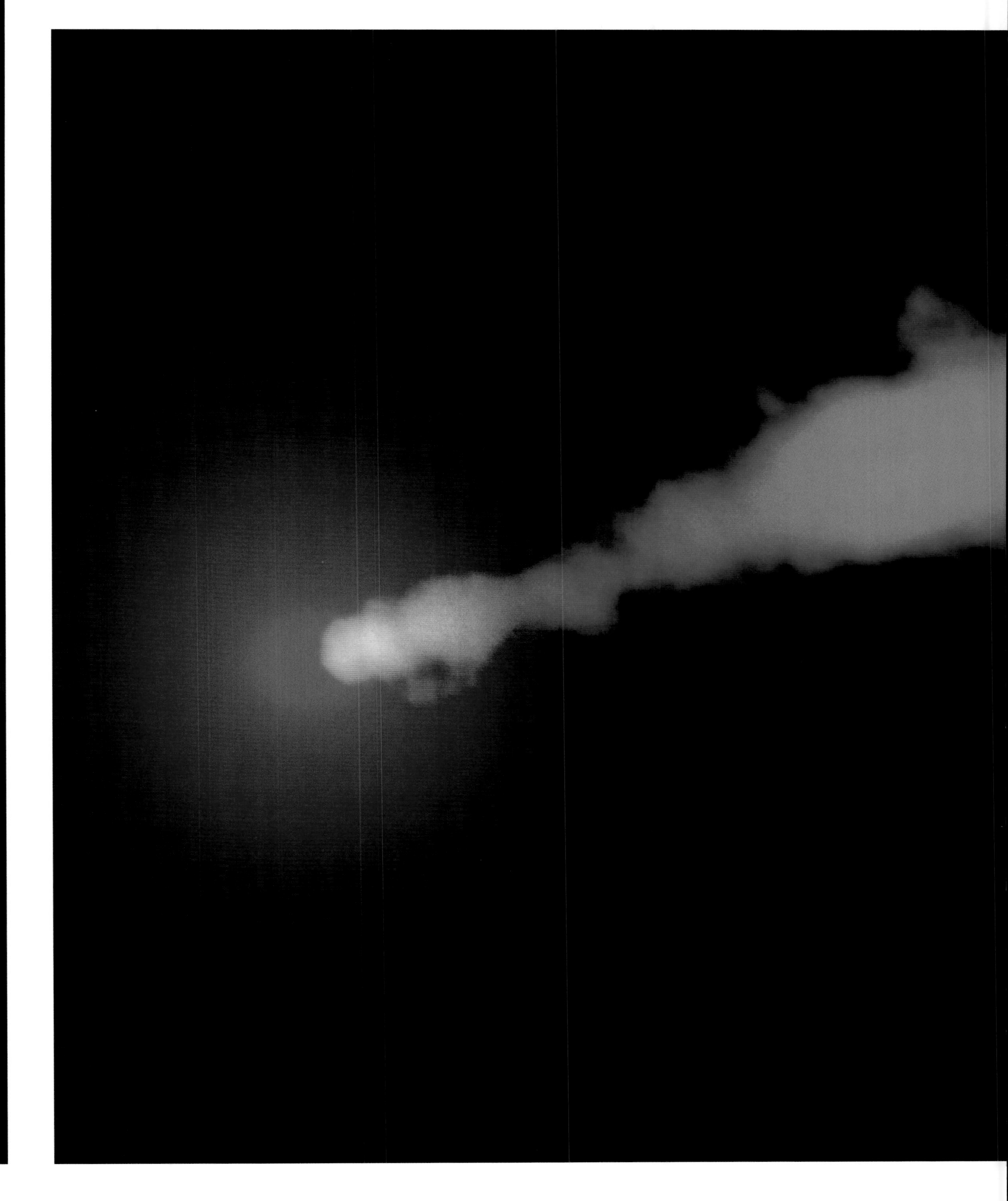

M87 "jet"

A COSMIC BLOWTORCH, 7,000 light-years long: M87's troubled center has given rise to this smoking gun, which blasts gas into space at velocities close to the speed of light. The jet was discovered at the Lick Observatory in California by Heber Curtis in 1918. Its eerie blue light is created by electrons tangled up in powerful magnetic fields, which are forced to emit synchrotron radiation. The gas is turbulent; it swirls and tumbles like water spewing out of a hose. But what a great view for the inhabitants of M87.

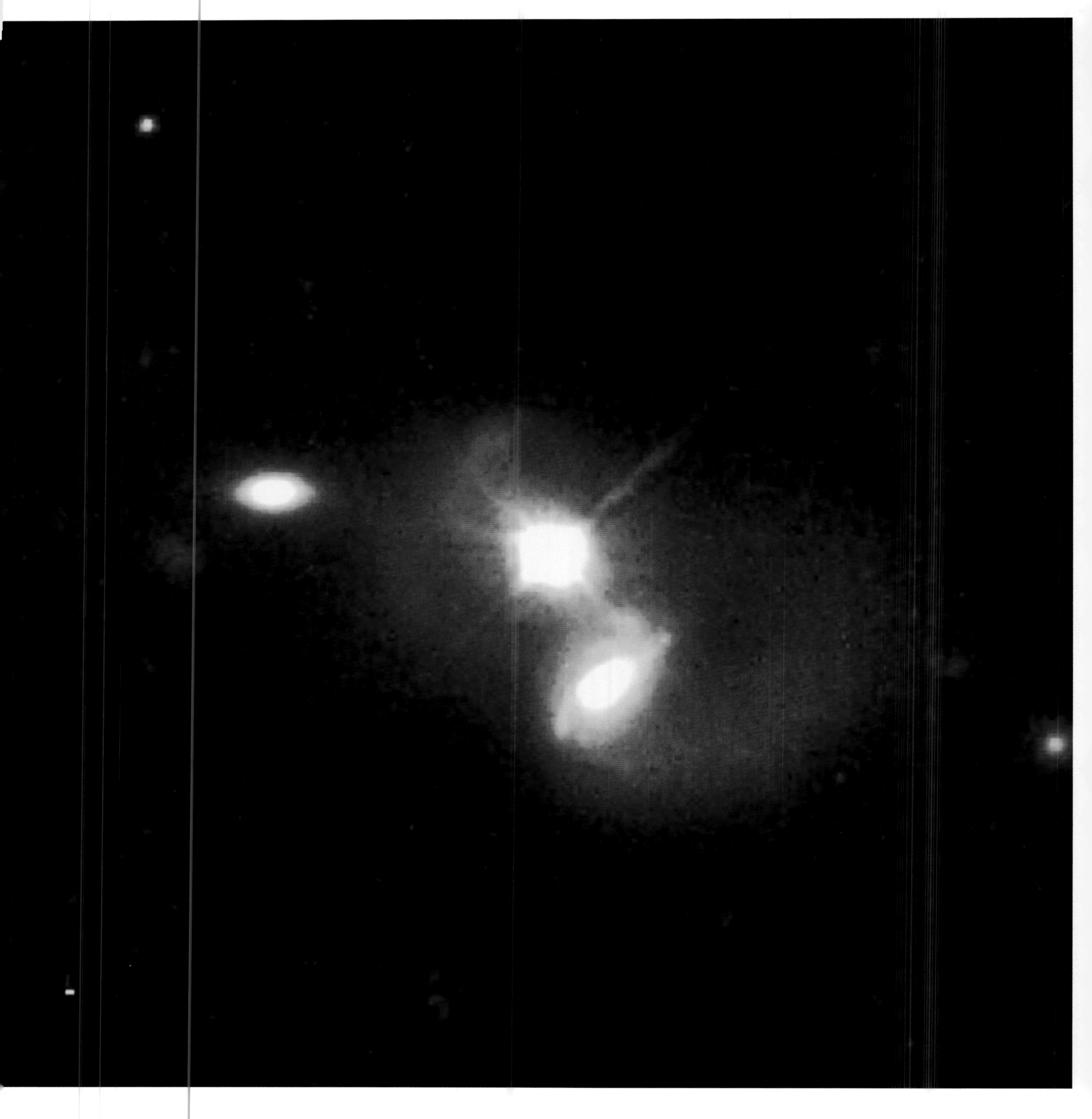

Quasar interacting with a companion galaxy

TWO GALAXIES IN A FATAL COSMIC EMBRACE: at the top, Quasar PG 1012+008 is about to eat its companion alive. Quasars were discovered by the Dutch American astronomer Maarten Schmidt in 1963, when he realized that the light from a "star" he was observing failed to match up to the fact that it was pouring out radio waves at an inappropriate rate. "Quasi-stellar radio-sources" rapidly became "quasars." These are the ultimate galaxies, powered by the biggest black holes at their centers.

Quasar Markarian 205

THE BEAUTIFUL SPIRAL GALAXY NGC 4319 curls up—like a celestial cat—in this image from the Hubble Space Telescope taken in October 2002. Other galaxies also speckle the frame. But the object on the top right is a quasar, Markarian 205. It looks close to the spiral galaxy, but it actually lies fourteen times farther away. The quasar appears brilliant, however, because of the violent radiation from its central accretion disk, where a massive black hole is devouring stars and gas.

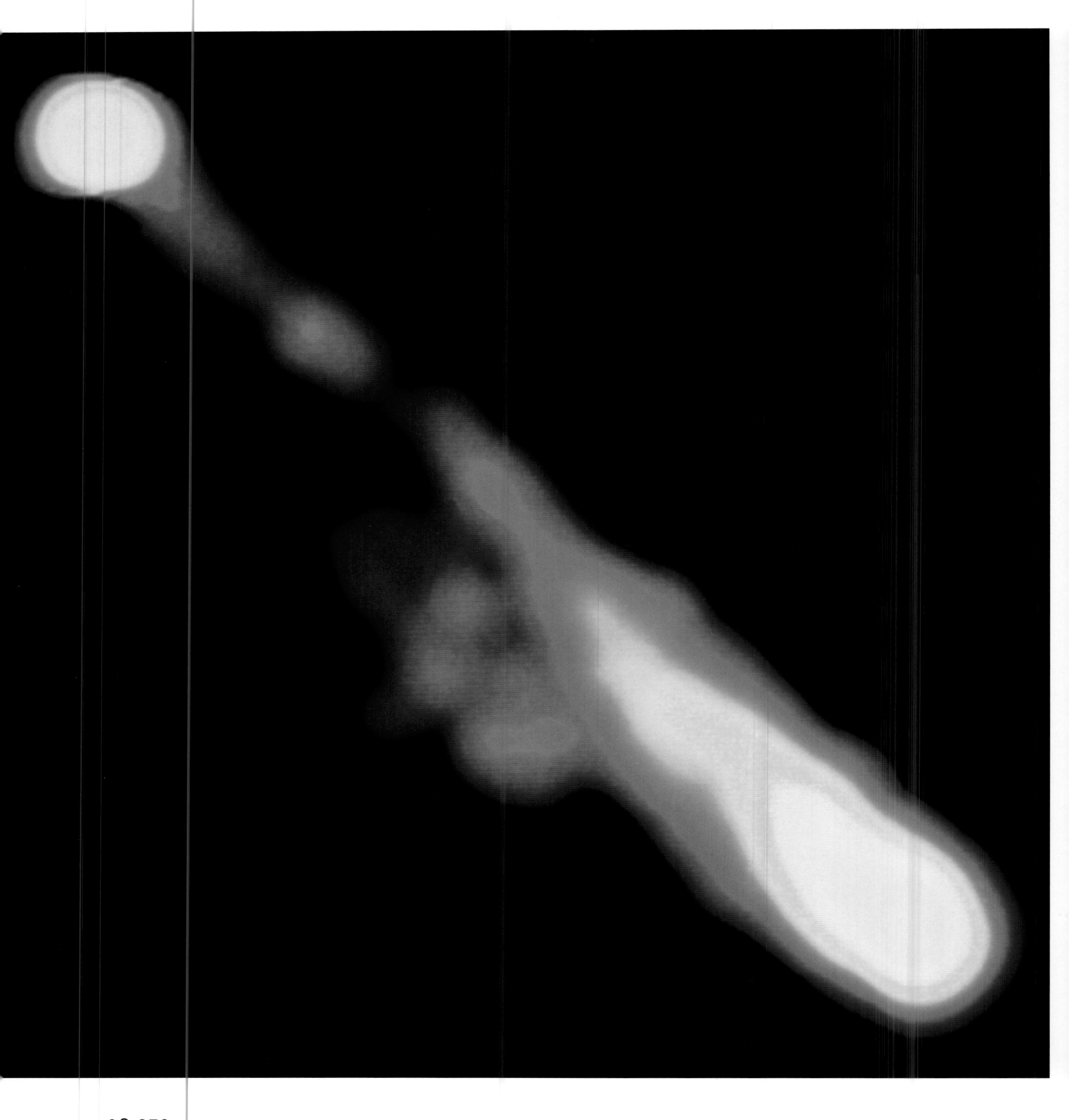

3C 273

WHEN RADIO ASTRONOMERS IN CAMBRIDGE published their third catalog of celestial radio beacons in 1959, they had no reason to suppose that number 273 would be significant. A few years later, though, they had to rethink this. 3C 273 turned out to be the brightest quasar in the sky—and a very unusual quasar to boot, with an unusual glowing "jet." The core of the quasar (top left) contains a black hole, while the jet is a thin beam of magnetism, stretching over 100,000 light-years into space. This powerhouse of energy lies way beyond the familiar galaxies, some 2,200 million light-years from Earth.

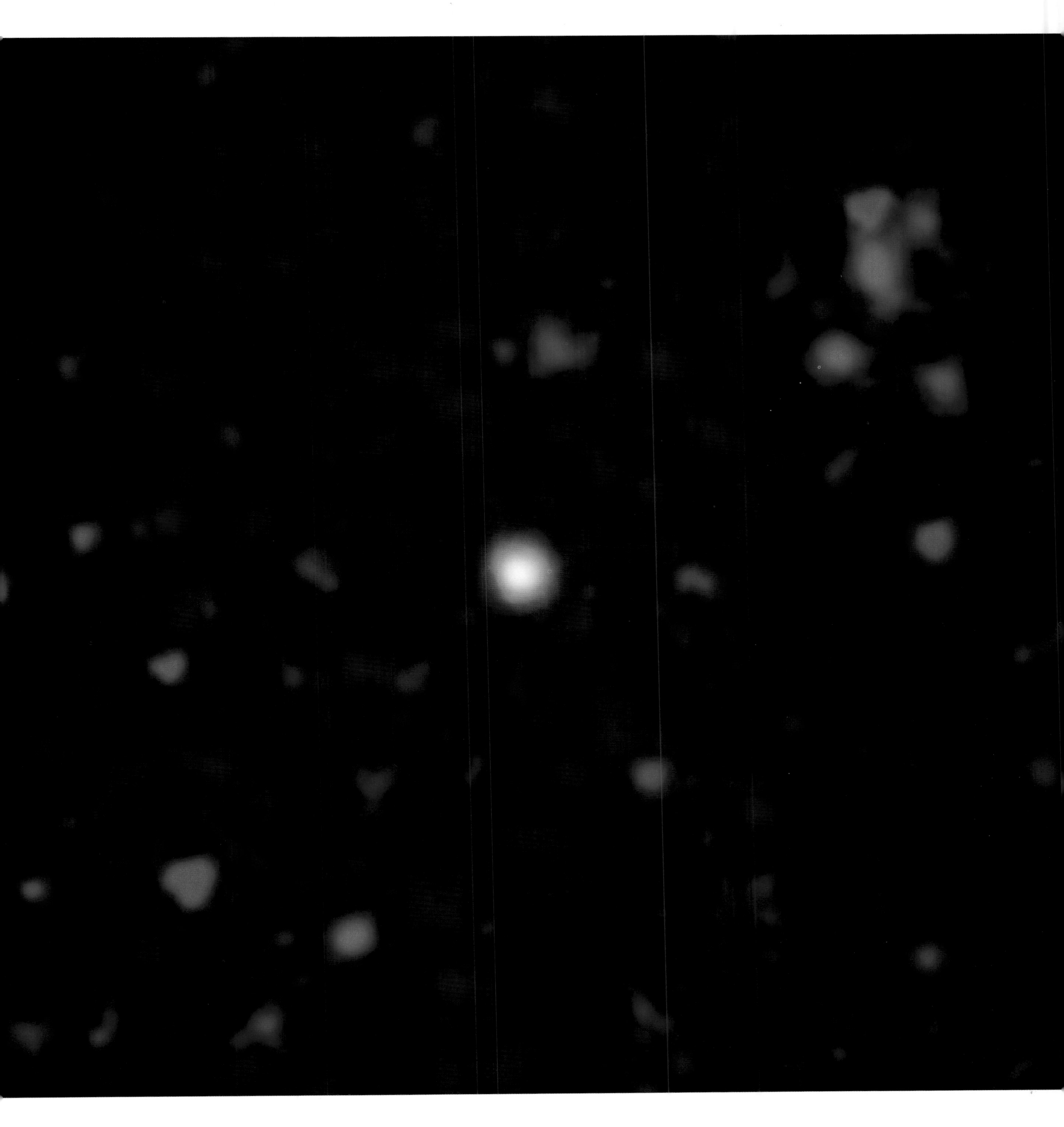

Quasar

BILLIONS OF LIGHT-YEARS AWAY in the depths of space, this quasar is not visible to even the world's largest telescopes. Cloaked in dust by the host galaxy, its blazing accretion disk can only be detected by the penetrating eye of the orbiting Spitzer Telescope—which finds celestial objects by the heat radiation that they emit. In this color-coded image, the blue, red, and green background tapestry reveals hundreds of galaxies living far away at the limits of the universe.

IN THE BEGINNING, THERE WAS NOTHING. It was a "nothing" so profound that it defies human comprehension. Space did not exist. Time did not flow. There was no matter, no radiation.

Then it all changed. From nothing, a tiny speck of light appeared. It was almost infinitely hot—and inside this fireball was the whole of space. With the creation of space came the birth of time: 13.7 billion years ago, the great cosmic clock began to tick. Ever since, our universe has been expanding.

There was no "before" the big bang. No one knows *why* it happened—except that it did. And the infant cosmos had another little trick up its sleeve. Like any young creature, it had more energy than it knew what to do with. Fractions of a second after forming, it embarked on a dramatic growth spurt. Had this "cosmic inflation" not taken place, the universe might have collapsed back on itself.

The energy in the primordial fireball was so concentrated that matter spontaneously started to appear. At this stage, atoms could not exist in the raging inferno. Instead, subatomic particles like quarks and bosons seethed and crashed into one another—sometimes building up a stable relationship, sometimes not.

After three minutes, the frenzied young universe had calmed down sufficiently to create the elements hydrogen and helium. But it would be over 200 million years before the first galaxies were born.

These came together from humble beginnings. Clumps of stars—like our present-day globular clusters—assembled themselves into bigger edifices. These galaxies started to swarm together in groups and clusters, creating the architecture of the universe as we know it today.

Meanwhile, stars continued to process the hydrogen fuel in their central cores, welding it into successively heavier elements. When these stars died, they enriched the spaces between the stars with these elements, including carbon, oxygen, and iron—making it possible for small solid planets to form and for life to be created.

This is the picture of our cosmos as we see it today, but it is only a partial picture. We know from the movement of galaxies in dense clusters that there is a lot more matter in the universe than meets the eye. Over 90 percent of the cosmos is invisible. It takes the form of mysterious "dark matter"—a catchy phrase, but one meaning that scientists have no clue what is out there.

And the picture grows curiouser and curiouser. Over the past few years—by establishing distances to remote supernovas—astronomers have discovered that our universe is not just expanding but *accelerating*. What is driving it? All we can surmise at the moment is that the cause must be some kind of negative force—"dark energy"—that is actively pushing the galaxies apart.

We live in a universe that will take a long time to comprehend. But *that* is the quest; *that* is what drives the human imagination. In the end, our far more intelligent descendants will regard themselves as privileged to inhabit such a beautiful and complex cosmos.

10 BIG BANG

Previous pages | Galaxy cluster: Abell 1060

A DAZZLING STAR IN THE MILKY WAY adorns our view of distant galaxies in the cluster Abell 1060. Galaxy clusters are part of the large-scale architecture of the universe, created in the big bang. Most of a cluster's mass, however, is not in the glittering galaxies but in some form of invisible "dark matter." What is it? And was it created in the big bang itself? This cluster, which spans about 10 million light-years, lies in the constellation Hydra.

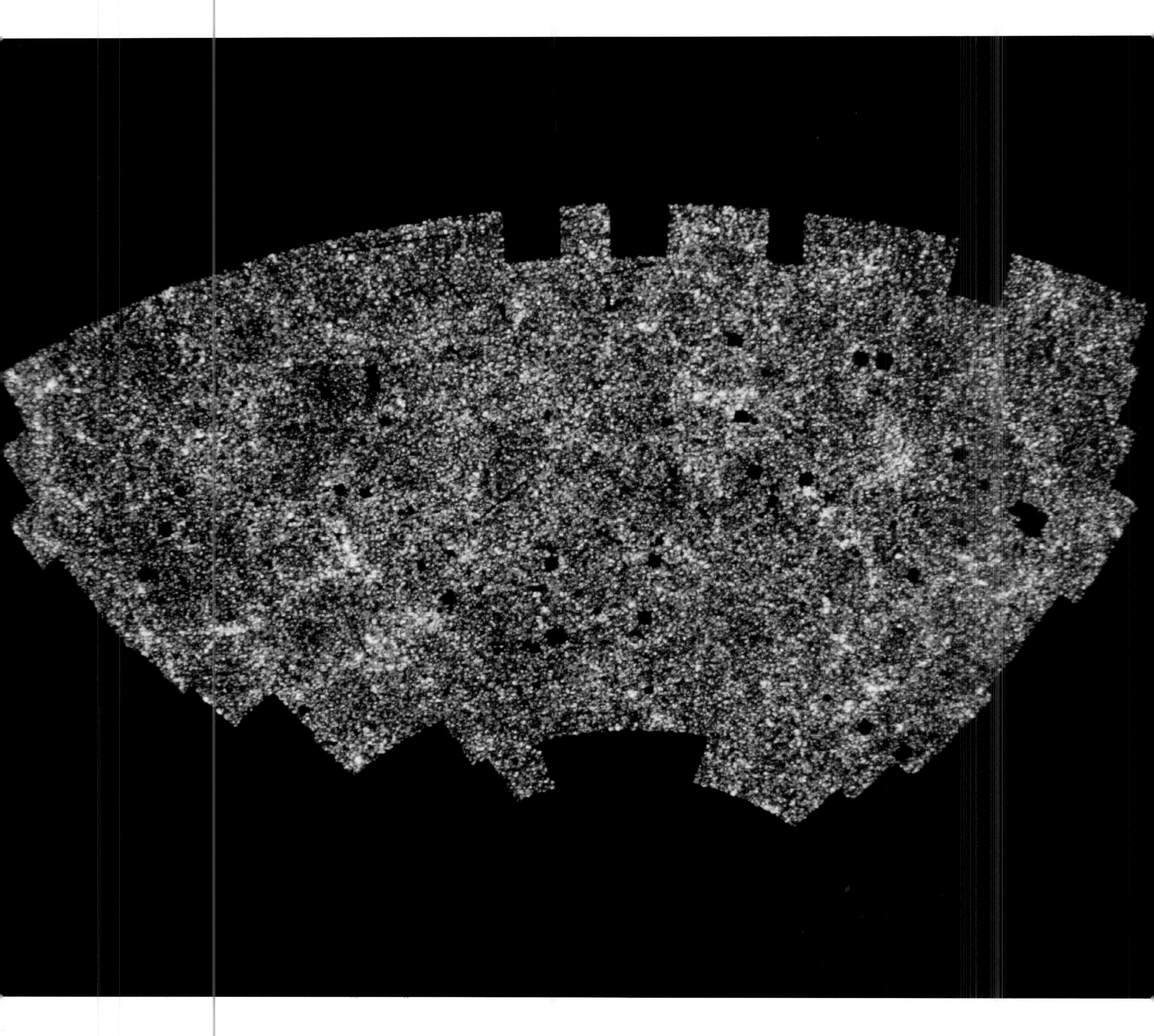

Map of galaxy distribution

LIKE A POPULATION MAP OF A COUNTRY ON EARTH, this view of one-tenth of the sky shows where galaxies reside. They clearly do not scatter at random. Galaxies live in conurbations—clusters and superclusters—linked into vast regions of urban sprawl, known as filaments and sheets. In between lie empty regions, over 100 million light-years across, that are devoid of galaxies. Theories of the big bang must explain why the universe has this "holey" structure, like a sponge or Swiss cheese.

Cosmic microwave background

THIS IS THE ULTIMATE BABY PICTURE—a view of the Universe in its earliest infancy. The WMAP satellite has peered deep into the cosmos to detect the faintest whispers of radiation from the birth of our universe. It has revealed an amazing image of the cloud of hot gases that erupted from the big bang. To put it in human terms: if we regard the universe today as middle-aged, then this view shows it at less than a day old. The emptier regions (blue) will grow into empty voids, while the denser regions of gas (yellow and red) will condense into swarms of galaxies.

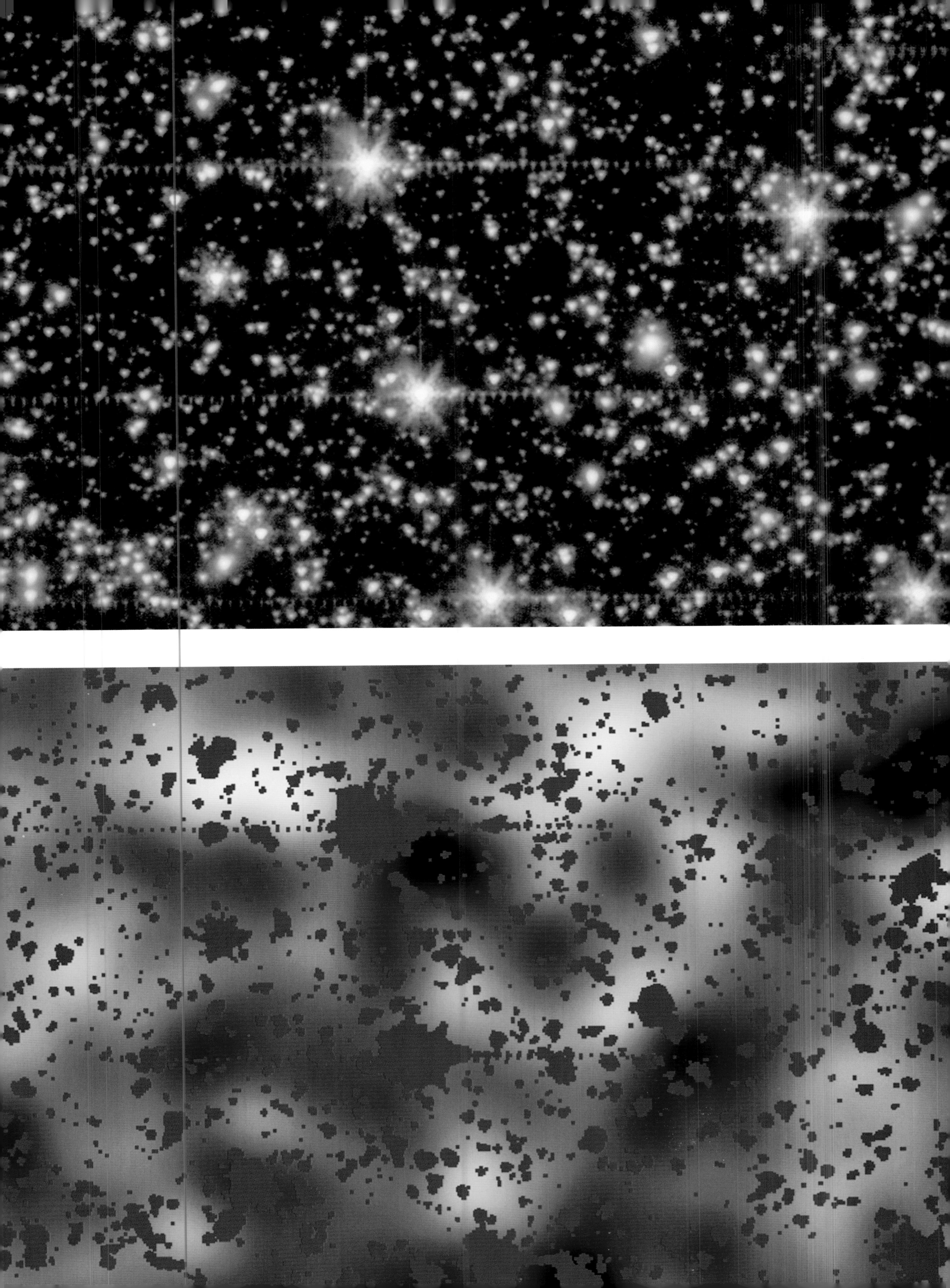

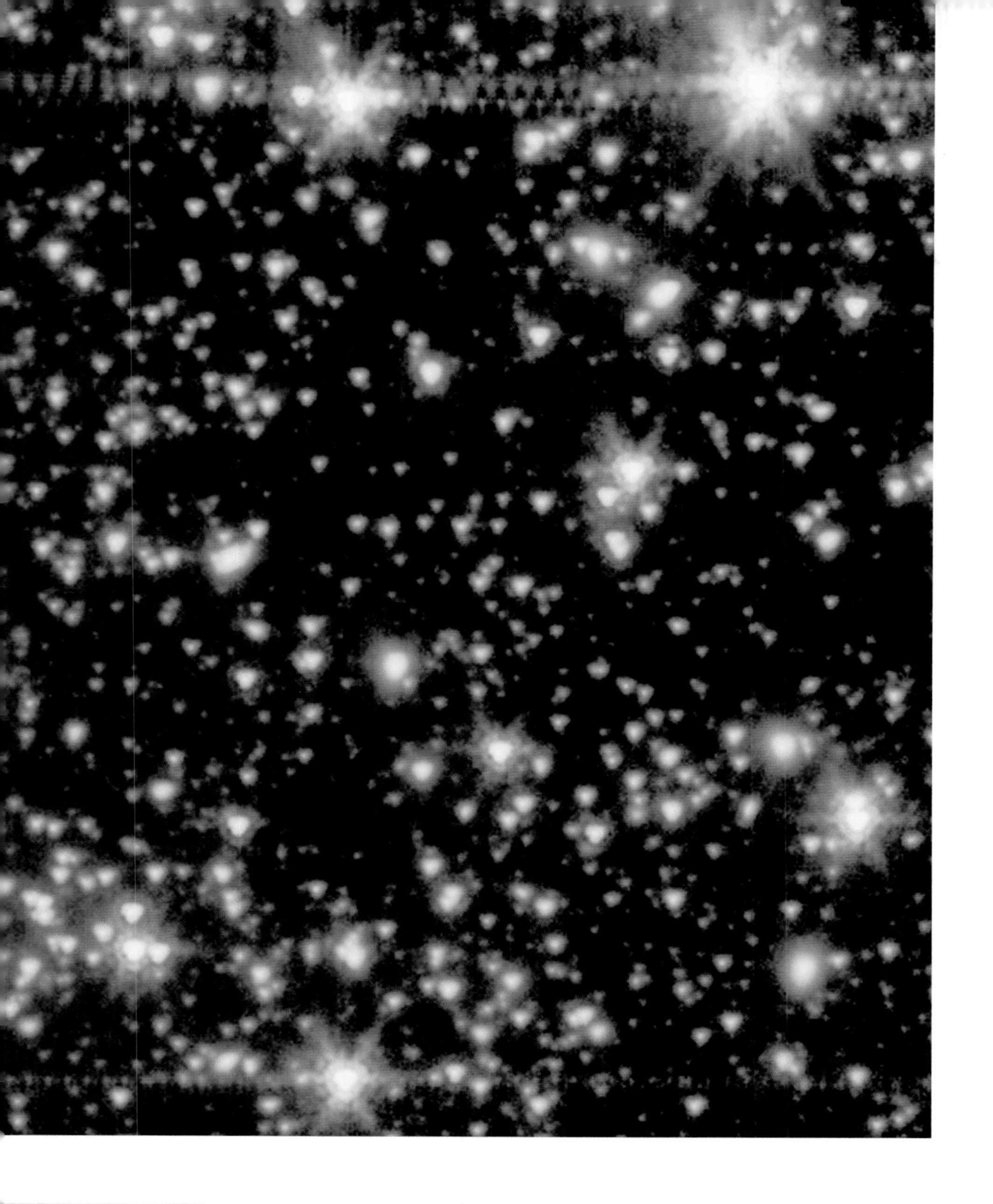

Spitzer image of distant stars

THE SPITZER SPACE TELESCOPE—the heat-seeking twin of the more famous Hubble—peered for ten hours into the depths of space to seek out the earliest stars in the universe. Its view of a tiny patch of sky in the constellation Draco reveals the most distant stars in our Milky Way and faint galaxies littering the farthest reaches of the cosmos. But its goal was to look further, into the dark patches between the stars and galaxies. Here, the Spitzer scientists hoped to pick out light that had been traveling for so long that it would reveal the earliest denizens of the universe . . .

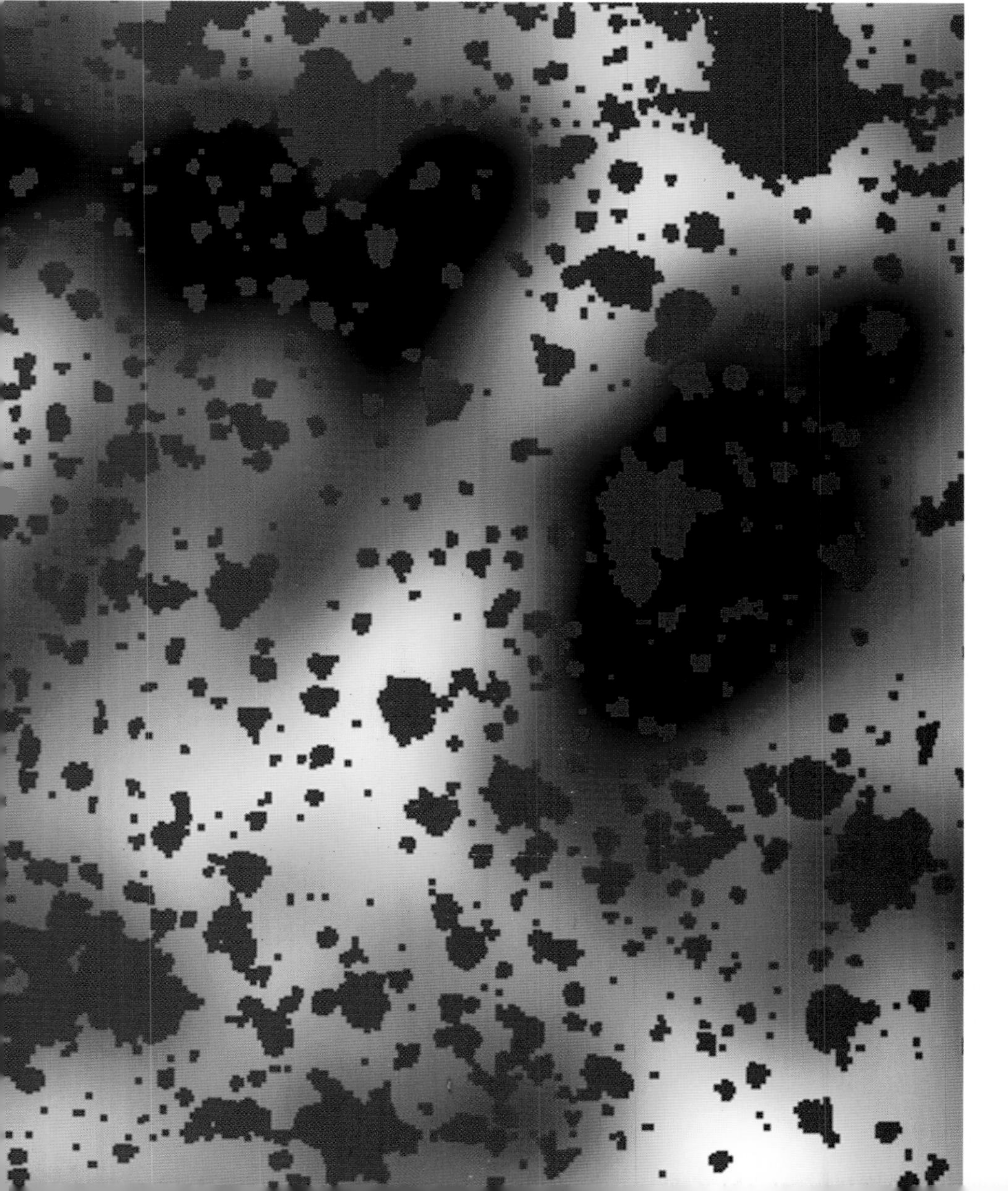

Light from the first stars

. . . AND WHEN THE EAGER SCIENTISTS blocked out the stars and galaxies, and turned up the brightness and contrast, they discovered that the apparently black background is in fact smoldering with infrared light. Here we see the glow of the first stars, born just 200 million years after the big bang (in human terms, when the universe was only six months old). These primeval stars were a hundred times heavier than our Sun and burned far more fiercely. They died in cataclysmic supernova explosions that seeded space with elements—like carbon, iron, and gold—that had never been seen in the universe before.

Globular cluster 47 Tucanae

BUILDING BLOCK OF THE GALAXIES: Globular cluster 47 Tucanae (affectionately called "47 Tuc") is a dense ball of stars easily visible in the southern hemisphere, close to the Small Magellanic Cloud in the sky. Over two hundred globular clusters orbit the Milky Way, and 47 Tuc—with its hundreds of thousands of stars—is the second brightest. Astronomers now believe that globular clusters gave birth to fledgling galaxies from the "bottom up," coming together in the millions in the distant past to create these huge cities of stars.

Globular cluster M3

GLOBULAR CLUSTERS ARE AMONG THE MOST ANCIENT inhabitants of the universe. Many were created some 12 billion years ago (47 Tuc is a relative youngster—a mere 10 billion years old!). M3, imaged here at the Kitt Peak National Observatory, is one of the largest and brightest: you can easily see it in the northern constellation of Canes Venatici through binoculars. Look at the scattering of ancient red stars—these are all on the way out. Globulars contain no gas, so they cannot regenerate themselves. M3 is home to half a million stars and lies 100,000 light-years away.

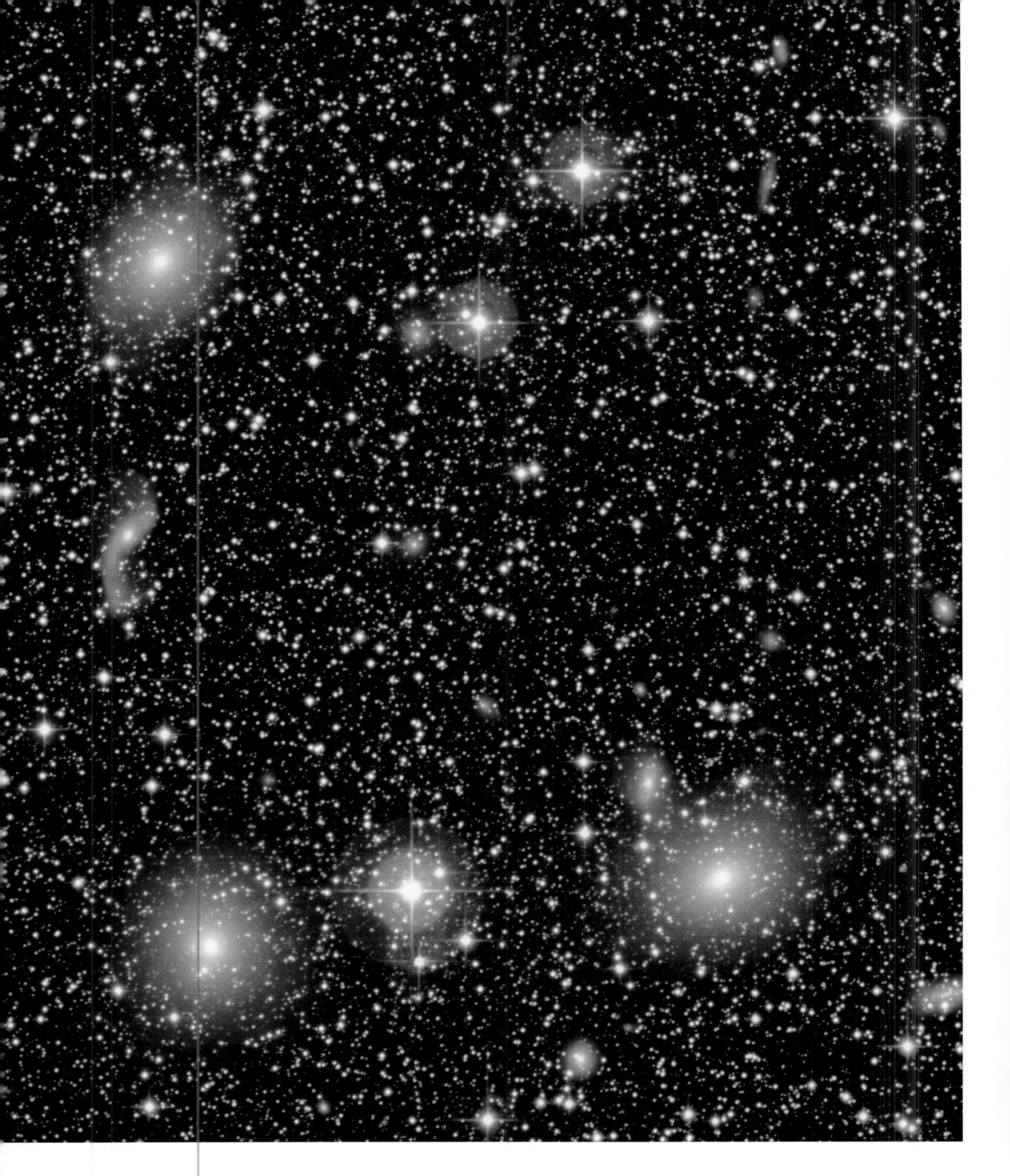

Galaxy cluster ACO 3627

COSMIC ARCHITECTURE WRIT *VERY* LARGE: This beautiful galaxy cluster lies in the direction of the southern constellation Centaurus. ACO 3627 forms the heart of a huge agglomeration of star cities 200 million light-years away. This supercluster—the Great Attractor—is tugging on the Milky Way, and it is even pulling the vast Virgo Galaxy Cluster toward it. Astronomers estimate that the total mass of the Great Attractor could be as high as 10,000 million, million Suns. The gravity of huge structures like this can disrupt the outward momentum of galaxies, initially caused by the big bang.

Stephan's Quintet

ANOTHER EXAMPLE OF COSMIC ARCHITECTURE, but here we see it up close and personal. And it is a reminder of how our universe might have been when the first galaxies were born, shortly after the big bang. At least two of the entangled galaxies of Stephan's Quintet—seen here in an image from the Hubble Telescope—have undergone hit-and-run accidents at high speeds. Hubble researcher Sarah Gallagher notes: "It's a phenomenon typical of the early universe, when encounters were much more common." But out of the wreckage, new life has emerged. The collision has led to the creation of at least one hundred new star clusters.

Gravitational lens 0024+1654

A COSMIC KALEIDOSCOPE IS REVEALED in this magnificent picture from the Hubble Space Telescope. Huddled in the center of the shot are the major members of a galaxy cluster that is 5 billion light-years from us. The elongated blue galaxies lie in the background, twice as far away. But look more carefully: five of the blue galaxies are identical. They are actually five images of the *same* galaxy. The gravity of the foreground cluster has focused the light from the blue galaxy so that its light reaches us from several different directions, like the multiple images in a kaleidoscope.

Lensed galaxies in a cluster

GRAVITY AGAIN CREATES THE LIGHT FANTASTIC when we view the galaxy cluster Abell 1689, named for astronomer George Abell, who was so fascinated by these cosmic giants that he created a catalog. Very distant galaxies appear as long, blue, curved arcs. But all the stars in all the galaxies in Abell 1689 do not have enough gravity to cause distortions *this* weird. A much stronger gravitational pull is at work. Observations like these are proving that the matter we see in the universe—galaxies, stars, and planets—constitutes only a minor player: the bulk of the cosmos is made of still-mysterious "dark matter."

INDEX

CREDITS

THE COPYRIGHT OF THE IMAGES USED IN THIS BOOK BELONGS TO THE FOLLOWING: 2–3, 164 NASA/ESA/JPL-CALTECH/J. HORA (CFA) AND C.R. O'DELL (VANDERBILT); 4 (TOP LEFT), 34 NASA/SCIENCE PHOTO LIBRARY; (TOP RIGHT), 53 NASA/SCIENCE PHOTO LIBRARY; (UPPER MIDDLE LEFT), 82 (TOP) NASA/SCIENCE PHOTO LIBRARY; (UPPER MIDDLE RIGHT), 118 NOAO/AURA/NSF/SCIENCE PHOTO LIBRARY; (CENTRE LEFT) 135 NASA, DONALD WALTER (SOUTH CAROLINA STATE UNIVERSITY), PAUL SCOWEN AND BRIAN MOORE (ARIZONA STATE UNIVERSITY); (CENTRE RIGHT), 170 NASA/ESA/STSCI/SCIENCE PHOTO LIBRARY; (LOWER MIDDLE LEFT), 210 NASA/SCIENCE PHOTO LIBRARY; (LOWER MIDDLE RIGHT), 235 NASA/SCIENCE PHOTO LIBRARY; (BOTTOM LEFT), 266 NASA/ ESA/ STSCI/SCIENCE PHOTO LIBRARY; (BOTTOM RIGHT), 283 NOAO/AURA/NSF/ SCIENCE PHOTO LIBRARY; 6, 150–151 NASA, ESA, M. ROBBERTO (SPACE TELESCOPE SCIENCE INSTITUTE/ESA) AND THE HUBBLE SPACE TELESCOPE ORION TREASURY PROJECT TEAM; 10–11 FRANK ZULLO/ SCIENCE PHOTO LIBRARY; 12 GEOFF TOMPKINSON/SCIENCE PHOTO LIBRARY; 13 WILLIAM ERVIN/SCIENCE PHOTO LIBRARY; 14 TONY CRADDOCK/SCIENCE PHOTO LIBRARY; 15 LARRY LANDOLFI/ SCIENCE PHOTO LIBRARY; 16 NASA/SCIENCE PHOTO LIBRARY; 17 NASA/SCIENCE PHOTO LIBRARY; 18 DAVID NUNUK/SCIENCE PHOTO LIBRARY; 19 GEORGE POST/SCIENCE PHOTO LIBRARY; 21 DAVID NUNUK/SCIENCE PHOTO LIBRARY; 22 NASA/SCIENCE PHOTO LIBRARY; 23 DR FRED ESPENAK/SCIENCE PHOTO LIBRARY; 24 KENT WOOD/SCIENCE PHOTO LIBRARY; 25 KENNETH W. FINK/SCIENCE PHOTO LIBRARY; 26 NASA/SCIENCE PHOTO LIBRARY; 27 JISAS/LOCKHEED/SCIENCE PHOTO LIBRARY; 28 EUROPEAN SPACE AGENCY/SCIENCE PHOTO LIBRARY; 29 SOHO/ESA/NASA/SCIENCE PHOTO LIBRARY; 30 NATIONAL OPTICAL ASTRONOMY OBSERVATORIES/SCIENCE PHOTO LIBRARY; 31 NATIONAL OPTICAL ASTRONOMY OBSERVATORIES/SCIENCE PHOTO LIBRARY; 32 NASA/SCIENCE PHOTO LIBRARY; 33 NASA/SCIENCE PHOTO LIBRARY; 35 NATIONAL OPTICAL ASTRONOMY OBSERVATORIES/SCIENCE PHOTO LIBRARY; 36 SOHO/ESA/NASA/ SCIENCE PHOTO LIBRARY; 37 NASA/SCIENCE PHOTO LIBRARY; 38–39 LARRY LANDOLFI/ SCIENCE PHOTO LIBRARY; 40 ROBIN SCAGELL/SCIENCE PHOTO LIBRARY; 41 DAVID NUNUK/SCIENCE PHOTO LIBRARY; 42–43 WALTER PACHOLKA, ASTROPICS/SCIENCE PHOTO LIBRARY; 44 DETLEV VAN RAVENSWAAY/SCIENCE PHOTO LIBRARY; 45 NASA/SCIENCE PHOTO LIBRARY; 46 DR FRED ESPENAK/SCIENCE PHOTO LIBRARY; 47 NASA/CUSTOM MEDICAL STOCK PHOTO/SCIENCE PHOTO LIBRARY; 48 ECKHARD SLAWIK/SCIENCE PHOTO LIBRARY; 49 ECKHARD SLAWIK/SCIENCE PHOTO LIBRARY; 50 G.ANTONIO MILANI/SCIENCE PHOTO LIBRARY; 51 DR FRED ESPENAK/SCIENCE PHOTO LIBRARY; 52 NASA/SCIENCE PHOTO LIBRARY; 54 NASA/SCIENCE PHOTO LIBRARY; 55 NASA/SCIENCE PHOTO LIBRARY; 56 NASA/SCIENCE PHOTO LIBRARY; 57 NASA/SCIENCE PHOTO LIBRARY; 58–59 REV. RONALD ROYER/SCIENCE PHOTO LIBRARY; 60 US GEOLOGICAL SURVEY/SCIENCE PHOTO LIBRARY; 61 NASA/SCIENCE PHOTO LIBRARY; 62 NASA/SCIENCE PHOTO LIBRARY; 63 PEKKA PARVIAINEN/SCIENCE PHOTO LIBRARY; 64 NASA/SCIENCE PHOTO LIBRARY; 65 NASA/SCIENCE PHOTO LIBRARY; 66–67 US GEOLOGICAL SURVEY/SCIENCE PHOTO LIBRARY; 67 NOVOSTI/SCIENCE PHOTO LIBRARY; 68 NASA/SCIENCE PHOTO LIBRARY; 69 NASA/SCIENCE PHOTO LIBRARY; 70 US GEOLOGICAL SURVEY/SCIENCE PHOTO LIBRARY; 71 NASA/SCIENCE PHOTO LIBRARY; 72 EUROPEAN SPACE AGENCY/SCIENCE PHOTO LIBRARY; 73 NASA/JPL/CORNELL/SCIENCE PHOTO LIBRARY; 74 EUROPEAN SPACE AGENCY/DLR/FU BERLIN (G. NEUKUM)/SCIENCE PHOTO LIBRARY; 75 NASA/SCIENCE PHOTO LIBRARY; 76 ESA/DLR/FU BERLIN (G. NEUKUM)/SCIENCE PHOTO LIBRARY; 77 ESA/DLR/FU BERLIN (G. NEUKUM)/SCIENCE PHOTO LIBRARY; 78 NASA/ESA/STSCI/SCIENCE PHOTO LIBRARY; 79 NASA/SCIENCE PHOTO LIBRARY; 80 NASA/SCIENCE PHOTO LIBRARY; 81 NASA/SCIENCE PHOTO LIBRARY; 82 (BOTTOM) NASA/JPL; 83 (TOP) NASA/SCIENCE PHOTO LIBRARY; (BOTTOM) NASA/SCIENCE PHOTO LIBRARY; 84 NASA/ESA/ STSCI/SCIENCE PHOTO LIBRARY; 85 NASA/SCIENCE PHOTO LIBRARY; 86 NASA/SCIENCE PHOTO LIBRARY; 87 NASA/SCIENCE PHOTO LIBRARY; 88 NASA/SCIENCE PHOTO LIBRARY; 89 NASA/JPL/SPACE SCIENCE INSTITUTE/SCIENCE PHOTO LIBRARY; 90 NASA/SCIENCE PHOTO LIBRARY; 91 NASA/SCIENCE PHOTO LIBRARY; 92 NASA/SCIENCE PHOTO LIBRARY; 93 NASA/SCIENCE PHOTO LIBRARY; 94 NASA/SCIENCE PHOTO LIBRARY; 95 NASA/ ESA/STSCI/SCIENCE PHOTO LIBRARY; 96 NASA/SCIENCE PHOTO LIBRARY; 97 NASA/SCIENCE PHOTO LIBRARY; 98 NASA/ESA/STSCI/SCIENCE PHOTO LIBRARY; 99 NASA/SCIENCE PHOTO LIBRARY; 100 NASA/ESA/STSCI/SCIENCE PHOTO LIBRARY; 101 NASA/ESA/ STSCI/SCIENCE PHOTO LIBRARY; 102–103 WALTER PACHOLKA, ASTROPICS/SCIENCE PHOTO LIBRARY; 104 DR FRED ESPENAK/SCIENCE PHOTO LIBRARY; 105 JOHN CHUMACK/SCIENCE PHOTO LIBRARY; 106 DAVID NUNUK/SCIENCE PHOTO LIBRARY; 107 FRANK ZULLO/SCIENCE PHOTO LIBRARY; 108 NASA/SCIENCE PHOTO LIBRARY; 109 DAVID PARKER/SCIENCE PHOTO LIBRARY; 110 NASA/ESA/STSCI/SCIENCE PHOTO LIBRARY; 111 NASA/ SCIENCE PHOTO LIBRARY; 112 NASA/ESA/STSCI/SCIENCE PHOTO LIBRARY; 113 MSSSO, ANU/SCIENCE PHOTO LIBRARY; 114 GORDON GARRADD/SCIENCE PHOTO LIBRARY; 115 REV. RONALD ROYER/SCIENCE PHOTO LIBRARY; 116 PEKKA PARVIAINEN/SCIENCE PHOTO LIBRARY; 117 SVEND AND CARL FREYTAG/ NOAO/AURA/NSF/SCIENCE PHOTO LIBRARY; 119 JOHN CHUMACK/SCIENCE PHOTO LIBRARY; 120 NASA/SCIENCE PHOTO LIBRARY; 121 JACK FINCH/SCIENCE PHOTO LIBRARY; 122 GORDON GARRADD/SCIENCE PHOTO LIBRARY; 123 JOHN CHUMACK/SCIENCE PHOTO LIBRARY; 124 DETLEV VAN RAVENSWAAY/ SCIENCE PHOTO LIBRARY; 125 NATIONAL OPTICAL ASTRONOMY OBSERVATORIES/SCIENCE PHOTO LIBRARY; 126 EUROPEAN SPACE AGENCY/SCIENCE PHOTO LIBRARY; 127 NASA/JPL-CALTECH/UMD/SCIENCE PHOTO LIBRARY; 128–129 NASA/ESA/STSCI/SCIENCE PHOTO LIBRARY; 130 CELESTIAL IMAGE CO./SCIENCE PHOTO LIBRARY; 131 J-C CUILLANDRE/CANADA-FRANCE-HAWAII TELESCOPE/SCIENCE PHOTO LIBRARY; 132 NASA/ESA/ STSCI/SCIENCE PHOTO LIBRARY; 133 NASA/ESA/STSCI/SCIENCE PHOTO LIBRARY; 134 J-C CUILLANDRE/CANADA-FRANCE-HAWAII TELESCOPE/SCIENCE PHOTO LIBRARY; 136 J-C CUILLANDRE/CANADA-FRANCE-HAWAII TELESCOPE/SCIENCE PHOTO LIBRARY; 137 NATIONAL OPTICAL ASTRONOMY OBSERVATORIES/SCIENCE PHOTO LIBRARY; 138 NATIONAL OPTICAL ASTRONOMY OBSERVATORIES/SCIENCE PHOTO LIBRARY; 139 NATIONAL OPTICAL ASTRONOMY OBSERVATORIES/SCIENCE PHOTO LIBRARY; 140 EUROPEAN SOUTHERN OBSERVATORY/SCIENCE PHOTO LIBRARY; 141 2MASS PROJECT/NASA/ SCIENCE PHOTO LIBRARY; 142 ROBERT GENDLER/SCIENCE PHOTO LIBRARY; 143 NASA/ ESA/STSCI/SCIENCE PHOTO LIBRARY; 144 RICHARD POWELL, DIGITIZED SKY SURVEY, PALOMAR OBSERVATORY; 145 M. SCHIRMER, T. ERBEN, M. LOMBARDI (IAEF BONN), EUROPEAN SOUTHERN OBSERVATORY; 146 CANADA-FRANCE-HAWAII TELESCOPE/JEAN- CHARLES CUILLANDRE/SCIENCE PHOTO LIBRARY; 147 2MASS PROJECT/NASA/SCIENCE PHOTO LIBRARY; 148 NASA/ESA/STSCI/SCIENCE PHOTO LIBRARY; 149 JASON WARE/SCIENCE PHOTO LIBRARY; 152 J-C CUILLANDRE/CANADA-FRANCE-HAWAII TELESCOPE/SCIENCE PHOTO LIBRARY; 153 NASA/ESA/STSCI/SCIENCE PHOTO LIBRARY; 154 NASA/SCIENCE PHOTO LIBRARY; 155 NASA/ESA/STSCI/SCIENCE PHOTO LIBRARY; 156 NATIONAL OPTICAL ASTRONOMY OBSERVATORIES/SCIENCE PHOTO LIBRARY; 157 NASA/JPL-CALTECH/R. A. GUTERMUTH (HARVARD-SMITHSONIAN CFA); 158–159 NASA/ESA/ STSCI/SCIENCE PHOTO LIBRARY; 160 NATIONAL OPTICAL ASTRONOMY OBSERVATORIES/ SCIENCE PHOTO LIBRARY; 161 NASA/ESA/STSCI/SCIENCE PHOTO LIBRARY; 162 NASA/ ESA/STSCI/SCIENCE PHOTO LIBRARY; 163 NASA/ESA/ STSCI/SCIENCE PHOTO LIBRARY; 165 NASA/ESA/STSCI/SCIENCE PHOTO LIBRARY; 166 NASA/ESA/STSCI/SCIENCE PHOTO LIBRARY; 167 EUROPEAN SOUTHERN OBSERVATORY/ SCIENCE PHOTO LIBRARY; 168 NOAO/AURA/NSF/SCIENCE PHOTO LIBRARY; 169 NASA/ESA/STSCI/SCIENCE PHOTO LIBRARY; 171 NASA/ESA/STSCI/SCIENCE PHOTO LIBRARY; 172 NASA/ESA/STSCI/SCIENCE PHOTO LIBRARY; 173 CELESTIAL IMAGE CO./SCIENCE PHOTO LIBRARY; 174–175 NASA/ ESA/STSCI/SCIENCE PHOTO LIBRARY; 176 NASA/ESA/STSCI/SCIENCE PHOTO LIBRARY; 177 ROBERT GENDLER/SCIENCE PHOTO LIBRARY; 178 CELESTIAL IMAGE CO./SCIENCE PHOTO LIBRARY; 179 NASA/ESA/STSCI/SCIENCE PHOTO LIBRARY; 180 CANADA-FRANCE-HAWAII TELESCOPE/JEAN-CHARLES CUILLANDRE/SCIENCE PHOTO LIBRARY; 181 ROBERT GENDLER/SCIENCE PHOTO LIBRARY; 182 NASA, ESA, J. HESTER (ARIZONA STATE UNIVERSITY); 183 NASA/ESA/STSCI/SCIENCE PHOTO LIBRARY; 184 NASA/SCIENCE PHOTO LIBRARY; 184 NASA/SCIENCE PHOTO LIBRARY; 186–187 JOHN CHUMACK/SCIENCE PHOTO LIBRARY; 188 J. BAUM & N. HENBEST/ SCIENCE PHOTO LIBRARY; 189 DENNIS DI CICCO/SCIENCE PHOTO LIBRARY; 190 LUKE DODD/SCIENCE PHOTO LIBRARY; 191 DENNIS DI CICCO, PETER ARNOLD INC./SCIENCE PHOTO LIBRARY; 192 ROYAL OBSERVATORY, EDINBURGH/AATB/SCIENCE PHOTO LIBRARY; 193 NASA/ESA/STSCI/SCIENCE PHOTO LIBRARY; 194 DR FRED ESPENAK/SCIENCE PHOTO LIBRARY; 195 FRANK ZULLO/SCIENCE PHOTO LIBRARY; 196 DR FRED ESPENAK/SCIENCE PHOTO LIBRARY; 197 ECKHARD SLAWIK/SCIENCE PHOTO LIBRARY; 198 DR LUKE DODD/ SCIENCE PHOTO LIBRARY; 199 ECKHARD SLAWIK/SCIENCE PHOTO LIBRARY; 200 ECKHARD SLAWIK/SCIENCE PHOTO LIBRARY; 201 DR FRED ESPENAK/SCIENCE PHOTO LIBRARY; 202 JERRY LODRIGUSS/SCIENCE PHOTO LIBRARY; 203 TONY & DAPHNE HALLAS/SCIENCE PHOTO LIBRARY; 204 JOHN CHUMACK/SCIENCE PHOTO LIBRARY; 205 JOHN CHUMACK/SCIENCE PHOTO LIBRARY; 206 JOHN CHUMACK/SCIENCE PHOTO LIBRARY; 207 TONY & DAPHNE HALLAS/SCIENCE PHOTO LIBRARY; 208 MOUNT STROMLO AND SIDING SPRING OBSERVATORIES/SCIENCE PHOTO LIBRARY; 209 ECKHARD SLAWIK/SCIENCE PHOTO LIBRARY; 211 NRAO/AUI/NSF/SCIENCE PHOTO LIBRARY; 212–213 DANIEL WANG, NORTHWESTERN UNIVERSITY/SCIENCE PHOTO LIBRARY; 214 MOUNT STROMLO AND SIDING SPRING OBSERVATORIES/SCIENCE PHOTO LIBRARY; 215 ROYAL OBSERVATORY, EDINBURGH/AATB/SCIENCE PHOTO LIBRARY; 216 CELESTIAL IMAGE CO./SCIENCE PHOTO LIBRARY; 217 NOAO/SCIENCE PHOTO LIBRARY; 218–219 NASA/ESA/STSCI/SCIENCE PHOTO LIBRARY; 220 NASA/ESA/STSCI/SCIENCE PHOTO LIBRARY; 221 EUROPEAN SOUTHERN OBSERVATORY/ SCIENCE PHOTO LIBRARY; 222 EUROPEAN SOUTHERN OBSERVATORY/ SCIENCE PHOTO LIBRARY; 223 NASA/ESA/STSCI/SCIENCE PHOTO LIBRARY; 224 EUROPEAN SOUTHERN OBSERVATORY/SCIENCE PHOTO LIBRARY; 225 NATIONAL OPTICAL ASTRONOMY OBSERVATORIES/COLOURED BY SCIENCE PHOTO LIBRARY; 226 EUROPEAN SOUTHERN OBSERVATORY/SCIENCE PHOTO LIBRARY; 227 NATIONAL OPTICAL ASTRONOMY OBSERVATORIES/COLOURED BY SCIENCE PHOTO LIBRARY; 228 NASA/THE HUBBLE HERITAGE TEAM (STSCI/AURA); 229 NASA/ESA/STSCI/SCIENCE PHOTO LIBRARY; 230 NASA/ESA/STSCI/SCIENCE PHOTO LIBRARY; 231 NATIONAL OPTICAL ASTRONOMY OBSERVATORIES/ SCIENCE PHOTO LIBRARY; 232 NASA/ESA/STSCI/SCIENCE PHOTO LIBRARY; 233 NASA/ESA/STSCI/ SCIENCE PHOTO LIBRARY; 234 NASA/ESA/STSCI/SCIENCE PHOTO LIBRARY; 236 NASA/ESA/THE HUBBLE HERITAGE TEAM (STSCI/AURA); 237 ROBERT GENDLER/SCIENCE PHOTO LIBRARY; 238 EUROPEAN SOUTHERN OBSERVATORY/ SCIENCE PHOTO LIBRARY; 239 EUROPEAN SOUTHERN OBSERVATORY/SCIENCE PHOTO LIBRARY; 240 NASA/ESA/STSCI/SCIENCE PHOTO LIBRARY; 241 EUROPEAN SOUTHERN OBSERVATORY/ SCIENCE PHOTO LIBRARY; 242 NASA/ESA/STSCI/SCIENCE PHOTO LIBRARY; 243 NASA/ESA/STSCI/SCIENCE PHOTO LIBRARY; 244 NASA/ESA/STSCI/SCIENCE PHOTO LIBRARY; 245 NASA/ESA/STSCI/SCIENCE PHOTO LIBRARY; 246 NASA/ESA/ STSCI/SCIENCE PHOTO LIBRARY; 247 NASA/ESA/STSCI/SCIENCE PHOTO LIBRARY; 248 NASA/ESA/ STSCI/SCIENCE PHOTO LIBRARY; 249 NASA/ESA/STSCI/SCIENCE PHOTO LIBRARY; 250 EUROPEAN SOUTHERN OBSERVATORY/SCIENCE PHOTO LIBRARY; 251 NOAO/AURA/ NSF/SCIENCE PHOTO LIBRARY; 252 TONY & DAPHNE HALLAS/SCIENCE PHOTO LIBRARY; 253 ROBERT GENDLER/SCIENCE PHOTO LIBRARY; 254 CELESTIAL IMAGE CO./SCIENCE PHOTO LIBRARY; 255 NOAO/SCIENCE PHOTO LIBRARY; 256 CANADA-FRANCE-HAWAII TELESCOPE/JEAN-CHARLES CUILLANDRE/SCIENCE PHOTO LIBRARY; 257 JOHN CHUMACK/SCIENCE PHOTO LIBRARY; 258–259 NASA/ESA/STSCI/SCIENCE PHOTO LIBRARY; 260 NASA/SCIENCE PHOTO LIBRARY; 261 CHANDRA X-RAY OBSERVATORY/ NASA/SCIENCE PHOTO LIBRARY; 262 CELESTIAL IMAGE CO./SCIENCE PHOTO LIBRARY; 263 J-C CUILLANDRE/CANADA-FRANCE-HAWAII TELESCOPE/SCIENCE PHOTO LIBRARY; 264 NASA/ESA/ STSCI/SCIENCE PHOTO LIBRARY; 265 NASA/ESA/STSCI/SCIENCE PHOTO LIBRARY; 267 NASA/CXC/IOA/A.FABIAN ET AL.; 268 NASA/ESA/STSCI/SCIENCE PHOTO LIBRARY; 269 DR RUDOLPH SCHILD/SCIENCE PHOTO LIBRARY; 270 NASA/ESA/STSCI/SCIENCE PHOTO LIBRARY; 271 NASA/ESA/STSCI/SCIENCE PHOTO LIBRARY; 272 JODRELL BANK/SCIENCE PHOTO LIBRARY; 273 NASA/SCIENCE PHOTO LIBRARY; 274–275 ROYAL OBSERVATORY, EDINBURGH/SCIENCE PHOTO LIBRARY; 276 PROF. GEORGE EFSTATHIOU/SCIENCE PHOTO LIBRARY; 277 NASA/WMAP SCIENCE TEAM/SCIENCE PHOTO LIBRARY; 278–279 (TOP) NASA/JPL-CALTECH/A. KASHLINSKY/SCIENCE PHOTO LIBRARY; (BOTTOM) NASA/JPL-CALTECH/A. KASHLINSKY/SCIENCE PHOTO LIBRARY; 280 NASA/ ESA/STSCI/SCIENCE PHOTO LIBRARY; 281 NOAO/AURA/NSF/SCIENCE PHOTO LIBRARY; 282 EUROPEAN SOUTHERN OBSERVATORY/SCIENCE PHOTO LIBRARY; 284 NASA/ ESA/STSCI/SCIENCE PHOTO LIBRARY; 285 NASA/ESA/STSCI/SCIENCE PHOTO LIBRARY.

Thunder Bay Press
An imprint of the Advantage Publishers Group
5880 Oberlin Drive San Diego, CA 92121-4794
www.thunderbaybooks.com

THUNDER BAY
P · R · E · S · S

All notations of errors or omissions should be addressed to Thunder Bay Press, Editorial Department, at the above address. All other correspondence (author inquiries, permissions) concerning the content of this book should be addressed to Cassell Illustrated, 2–4 Heron Quays, Canary Wharf, London, E14 4JP, London.

Library of Congress Cataloging in-Publication Data available on request.

ISBN-13: 978-1-59223-699-2
ISBN-10: 1-59223-699-5

Printed in China
1 2 3 4 5 10 09 08 07 06

Commissioning Editor: Karen Dolan
Editor: Katie Hewett
Design: Austin Taylor
Index: Bill Johncocks